AUTHORS OF SCIENTIFIC NAMES
IN PTERIDOPHYTA

AUTHORS OF SCIENTIFIC NAMES IN PTERIDOPHYTA

A list of authors of names of ferns and fern allies with recommended standard forms of their names including abbreviations

Compiled by

RODOLFO E.G. PICHI SERMOLLI

Collaborators:

MARIA PAOLA BIZZARRI
for bibliographical research, typesetting and proof reading

KUNG-HSIA SHING and XIAN-CHUN ZHANG
for biographical data and spelling of names of Chinese authors

ROYAL BOTANIC GARDENS, KEW
1996

ISBN 0 947643 90 7

Typeset by
R.E.G. Pichi Sermolli and Maria Paola Bizzarri

On the cover:

A fine form of *Phyllitis sagittata* (DC.) Guinea et Heywood which grew in the sixteenth century on some ancient monuments of Rome. First described and illustrated by C. Clusius with the name of «Hemionitis peregrina» (1576) and later by C. Durante (1585), R. Morison (1699) and A. Fiori (1943). Reproduced (half the size of the original) from Morison's Pl. Hist. Univ. 3: sect. 14. tab. 1. 1699.

Printed and bound in Italy by Litografia Europa, La Spezia

PREFACE

In 1954, at the Eighth International Botanical Congress in Paris I proposed to undertake the preparation of a further supplement to Christensen's *Index Filicum* and at the same time I suggested that an International Committee be appointed to prepare it. My proposal was discussed at a special meeting of the pteridologists present at the Congress and was unanimously accepted. One year later the International Association for Plant Taxonomy agreed to sponsor this enterprise, and in 1955 the Executive Committee appointed the 'Index Filicum Committee' consisting of six members, and I was designated to act as its Secretary.

From 1934 onwards, F.Ballard continued preliminary work at Kew towards the continuation of the *Index Filicum,* and had assembled some hundreds of cards; these, and the cards I had prepared in 1952-54, became the foundation of the card-index for the Fourth Supplement. But this was largely incomplete and I decided to undertake the scanning, volume by volume, of many periodicals (167 in all) and books in which *novitates* concerning ferns might be found. I also invited the Committee members to scan the periodicals and books which I had been unable to examine. Finally I made a revision of the basionyms of the new combinations cited in the Fourth Supplement of the *Index Filicum.* It was published at the end of January of 1965.

During the preparation of this Fourth Supplement (Christensen had published the original Index and three Supplements, the last in 1934), I realized that in the past the name of an author had been quoted either in full or abbreviated, with different abbreviations, and that sometimes the same abbreviation had been used for different authors. These observations and the wish to know something about the life and the activity of various pteridologists led me to undertake some biographical research on the authors who had established new taxa, new names, new combinations, new hybrids, or other *novitates* in Pteridophyta.

My research was laborious, but in a reasonably short space of time I was able to prepare a list of fern authors consisting of their names with pertinent abbreviations, dates of birth and, when applicable, of death. It was later published as an appendix (pp.351-361) to the Fourth Supplement of the *Index Filicum.*

After the publication of this Supplement the Index Filicum Committee, in 1968, was suppressed and the preparation was taken over by the Royal Botanic Gardens, Kew. It culminated in the publication in 1985 of the Fifth Supplement prepared by F.M.Jarrett and her collaborators. It provides an index to the names of ferns and fern allies at all ranks from family to species, published between 1961 and 1975.

During these years and in the following ones I continued my research on the authors of names in the Pteridophytes particularly during the time in which I was preparing my works on family names of Pteridophyta (1970), the classification of the higher taxa of Filicopsida (1973) and the *Tentamen* on the genera of Pteridophyta (1977). My attention was also turned to the authors of names of fern allies, as well as of hybrids and infraspecific taxa.

In 1980 the Royal Botanic Gardens, Kew published a *Draft Index of Author Abbreviations* for flowering plants compiled under the supervision of R.D.Meikle, and in 1985 it promoted the preparation of a revised version of the *Draft* and decided to include in it the authors of names of all groups of the plant kingdom. With this intent, late in the same year, this arduous enterprise was entrusted to an international 'Working Party' of many specialists in the various groups of plants. It operated for six years at Kew under the supervision of R.K.Brummitt acting as Convener and Secretary, with the assistence from 1989 to 1992 of C.E. Powell.

Firstly the Working Party was engaged in establishing a set of principles to be followed in compiling the lists of authors of the main taxonomic groups, and later in preparing and examining them in detail. Some preliminary work on ferns authors had been done at Kew by B.S.Parris. However, since I had published the Index of authors in Supplement four of the *Index Filicum*, I was invited to compile the list of the authors of names in Pteridophytes.

A consolidated list of the authors of all groups of plants, prepared in accordance with the principles established by the Working Party and including the recommended 'standard forms' of their names, was accomplished in 1991. In the following year it was edited by R.K.Brummitt and C.E.Powell in the form of a book with the title *Authors of Plant Names* published by the Royal Botanic Gardens, Kew.

After the publication of this book I decided to publish one similar to it in style and lay-out but limited to the authors of names in Pteridophyta. Thus, I undertook again my research with the main aims of ascertaining the forenames and the biographical data of the authors which had remained incomplete, and finding out the authors, particularly those of infraspecific names and hybrids, which had escaped my previous research, and also those who had established some *novitates* after 1991.

The results of my past and recent research are assembled in this book which is published, as that of Brummitt & Powell, by the Royal Botanic Gardens, Kew, to which I express my most sincere gratitude for having done me the honour to publish it under their auspices.

This work is devoted to all botanists who have contributed to the advancement of our knowledge of ferns and fern allies. I hope that it will prove to be useful as a guide for the citation of authors and as a reference book for all who are interested in the study of these fascinating plants.

INTRODUCTION

As I have mentioned in the Preface, R.K.Brummitt & C.E.Powell's book is mainly the result of the research of a Working Party consisting of many specialists in the different groups of the plant kingdom who between 1985 and 1991 collaborated to the realization of this enterprise which culminated in the publication in 1992 of the book *Authors of Plant Names.*

In the introduction to their book the authors point out its origin and scope and give a detailed report on the activity of the Working Party, especially for the compilation of data on the authors, for the solution of the problems of spelling and manner of citing some particular surnames, and finally for the definition of the principles followed in establishing the recommended 'standard forms' of the authors' names.

ACTIVITY OF THE WORKING PARTY

I consider it superfluous to deal again with the research carried out by the Working Party and with the list of the collaborators whose names are given by Brummitt & Powell in the acknowledgements of their book. As regards the biographical data, the manner of treating Cyrillic, Chinese and Ethiopian names, the names of Spanish and Portuguese origin, the citation of the prefixes in Dutch, Belgian, South African names and the German prefix 'von', I have fundamentally accepted the decisions and suggestions of the Working Party as reported in *A.P.N.*. However, some comments which emerged from my experience while preparing this book, will be given below when dealing with its compilation.

During its activity, the Working Party devoted its attention more to the 'standard forms' than to any other subject, and at last it laid down a set of principles, of which only very few were considered as absolute rules. I think it convenient and useful that they be known to all interested in pteridological research, both for understanding the reason for the selection of some 'standard forms' and for suggesting the way to establish those for future authors. However, before dealing with these principles I wish to give a definition of the 'standard form'. According to Brummitt & Powell (p.9) "The 'standard form' of a name is a surname, or an abbreviation of it (e.g. Adans. for Adanson), or rarely a contraction of it (e.g. Michx. for Michaux), with or without initials or other distinguishing appendages (see principles 7-9 below).".

The principles established by the Working Party are reported by Brummitt & Powell in their book on pages 9-12 of the introduction, and from them I list below all these principles, often quoting them verbatim. However, I have omitted some occasional comments and replaced some examples concerning authors of spermatophytes with others of pteridophytes.

These principles are 14. The first four and the 12th are considered as absolute rules and must be applied rigidly. The remaining ones are not absolutely binding. Principles no.6 to 14 are applied only to names of authors not appearing in the second edition of *Taxonomic Literature* by Stafleu & Cowan (1976-1988), well known and mostly quoted as *TL-2*.

1. *Script.* Names are given in Roman characters. When originally written in different characters they must be transliterated.

2. *Uniqueness.* Every 'standard form' must be unique to one person. Exceptions to this principle may be made for the persons who have published under completely different names during their lifetime (see below no.11).

3. *Uniform treatment of names.* The same surname (i.e. identical spelling) must always be given in the same form (e.g. Meyer is always abbreviated to Mey.), unless it is part of a compound name (see below, no.14), and different surnames must not be given in the same form (if Brown is abbreviated to Br. then Browne must not also be abbreviated to Br.).

4. *Full-stops and accents.* All abbreviations and contractions are terminated by a full-stop (e.g. Adans., Fern.Areces, Michx.) but the full-stop does not make a standard form different from the same spelling without a full stop (e.g. Lam. for Lamarck and Lam for H.J.Lam would be treated as homonymous, and initials are required for the latter's standard form). Similarly names differing only by presence or absence of an accent (e.g. Leonard and Léonard) or a diaeresis (Love and Löve) or an apostrophe (Ohara and O'Hara) are treated as homonymous.

5. *TL-2 as a standard.* The Working Party strongly favoured accepting the standard forms recommended there. This has mostly been done, but in some cases *TL-2* has not been followed. The reasons for diverging from *TL-2* are given by Brummitt & Powell; one of these reasons regards the cases in which the abbreviations adopted in *TL-2* conflict with particularly well established abbreviations used elsewhere, such as Copel. (instead of Cop.) which is widely adopted in pteridological literature for E.B.Copeland. In the Appendix to the present book I have listed all the standard forms of authors of pteridophyte names used in *TL-2* which differ from those adopted in *A.P.N.* and in the present book.

6. *Surname only.* A surname alone, or its abbreviation or contraction, is adopted as the standard form if it is applicable to only one author in the list. A surname alone, or its abbreviation or contraction, is usually also adopted for one of a number of authors with the same surname (see below no.7).

7. *Initials.* Persons with identical surnames are distinguished by use of initials of forenames, except as in the following no.8 & 9. Usually the earliest born is given without initials and all later ones with initials, but in some cases a better known later author may be given without initials and all others, including the earliest, with initials. Occasionally, when all having the same surname are more or less contemporary and equally well known, all may be given initials. Where initials are required: a) if the author has one forename, then the one initial is given; b) if the author has two forenames, both initials are given,

except in occasional cases where an author consistently omits one initial in authorship of his/her publications (as distinct from authorship of plant names), e.g P.Taylor, H.Rob., B.Nord. for P.G. Taylor, H.E. Robinson and R.B. Nordenstam who consistently publish their work using only one of their forenames; c) if the author has three forenames, three initials are given unless he/she has a clear preference for using only one and no ambiguity arises (W.D.J.Koch is given all three initials to avoid confusion with Walo Koch); if the author has more than three forenames, an *ad hoc* decision is taken. *It is recommended that no spaces be left after full stops.*

8. *Abbreviated or full forenames.* When two authors have identical surnames and initials, full or abbreviated forenames may be used. One author may be given with only surname, or initial(s) plus surname, and the other with fuller name, or both may be given fuller names, [e.g. Thomas Hogg (1777-1855) and Thomas Hogg (1820-1892) are distinguished as Hogg and T.Hogg respectively, while Walter Jones and William Jones have the standard forms Walt. Jones and Wm.Jones respectively].

9. *Suffixes.* In a few cases persons with identical surnames may be distinguished by a suffix instead of, or in addition to, initials. In well known cases of father and son, the son may be distinguished by 'f.', an abbreviation of 'filius'. Where this follows an abbreviated name with a full-stop, *it is recommended that no space be left between full-stop and f.* (e.g. Rech.f. for Rechinger the son). Tradition may also allow a different suffix, such as 'Arg.' for "Argoviensis" (of Aargau), 'Berol.' for "Berolinensis" (of Berlin) and 'Hal.' for "Halensis" (of Halle) in Müll.Arg., Müll.Berol. and Müll.Hal. respectively. In a few cases where different authors have identical surname and forename we have used the suffixes 'bis' for the second and 'ter' for the third, as in R.Br., R.Br. bis and R.Br. ter for the three Robert Browns.

10. *Variant names for the same person.* Except as noted in the following no.11, one person is always given the same standard form, even though he/she may have modified the spelling of the surname during his/her lifetime, or different transliterations may exist, or a compounding form may have been adopted during their lifetime. For example, Meisner adopted the spelling Meissner later in life but is always given the same standard form, Meisn. The variant transliterations Tsvelev, Tsvelov, Tzvelev and Tzvelov have all appeared for the same person, whose recommended standard form Tzvelev should be used in all cases. Alan Radcliffe Smith originally published plant names as A.R.Smith, but later modified his name to Radcliffe-Smith; the standard form Radcl.-Sm. should be used for both.

11. *Different names for the same person.* When a person has published under completely different names at different times, different standard forms may be used for the same person. For example, Doris Alma Goy has published both under her maiden surname Goy and her married surname Smith. Both may be used in the standard forms: Goy and D.A.Sm.. However, where one person has simultaneously used alternative names, such as Brother Alain who also uses the names Liogier, Enrique Eugenio, and others, a single name has been adopted (in this case Alain).

12. *Where to abbreviate*. Names are never abbreviated before a consonant. *This is an absolute rule*.

13. *How many letters to save in standard forms*.

a) Names are usually not abbreviated unless more than two letters are eliminated and replaced by a full-stop, unless recommended in *TL-2*.

b) Names of authors living before the 20th century are more likely to be abbreviated than later ones, and names of authors in the later 20th century tend to be given in full.

c) Where no strong tradition exists, names of 8 letters or fewer are not abbreviated, names of 9 letters are abbreviated if more than 3 letters are eliminated (e.g Verdc. for Verdcourt), and names of 10 letters or more are usually abbreviated.

d) Other things being equal, if an abbreviation is made, 2-syllable names are abbreviated to one syllable, 3-syllable names are most likely to be abbreviated to one (e.g. Beccari to Becc.), and names of 4 or more syllables are most often abbreviated to two syllables (e.g. Petitmengin to Petitm.). Many Japanese names have four short syllables, and we have tried to be consistent in abbreviating them to two syllables (e.g. Kitagawa to Kitag.) except where ambiguity would occur, or except where the name has only seven letters.

e) Application of these principles may depend on whether initials are also needed in standard forms of well known authors. Decisions may also be affected when an author is known to be commonly cited as a joint author with somebody else.

14. *Compound names*. Compound names, whether hyphenated or not, may be treated as special cases, and some principles given above may be discounted. In order to keep the standard form as short as possible, the principles of how many letters should be saved (see above no.13) may often be broken. Similarly the principle that the same name must always be given in the same way (see above no.13) is over-ridden in compound names, so that, for example, although Gonzales and Jenkins are given in full when they stand on their own, Gonzales Albo and Fraser-Jenkins are abbreviated to Gonz.Albo and Fraser-Jenk. respectively. *It is recommended again that no space be left after the full-stop*.

COMPILATION OF THE LIST

As mentioned in the Preface, the compilation of this list dates back to the years after 1955 when I was preparing the Fourth Supplement of the *Index Filicum*. To fulfill this task I was compelled to scan a great number of periodicals, floras, monographs and other books published after 1933. This research was very long and tedious, but the bulk of the data assembled at that time allowed me to prepare also the list of the authors of fern names. It was published as an appendix to the Fourth Supplement of the *Index Filicum* in 1965. In the following years I continued my bibliographical investigations and in 1991 I completed a second list of authors of names in Pteridophyta. It provided the data on these authors cited in Brummitt & Powell's book *"Authors of Plant Names"* published in 1992.

After the publication of this book I undertook again my research with the main scope of completing the data on some authors, particularly their forenames and biographical data, and to trace the names of other authors, especially of infraspecific taxa, who had escaped my previous investigations. It was carried on by bibliographical and biographical research and by a direct correspondence both with living authors and with the institutions where deceased authors had carried on their activity.

The bibliographical research was devoted to the scanning of books and periodicals published from 1991 to 1995, and of some publications I had not been able to examine in the previous years. Particularly fruitful was a period of about twenty days I spent in the Kew library in 1995 and other short visits made to the Central Libraries of the Natural History Museum of London and to that of Paris Museum d'Histoire Naturelle.

The research by correspondence, although expensive, was very useful and allowed me to complete many data and at the same time to check, and sometimes to correct, some forenames and years of birth and death. However, in spite of it, the data on various authors remained incomplete, and are still so because about twenty-five per cent of my letters addressed to living authors remained, unfortunately, unanswered.

The present book mainly consists of a list of the authors of scientific names in Pteridophyta and of an Appendix with the list of the abbreviations of authors' names adopted in past and recent publications, which differ from the standard forms adopted in *A.P.N.* and here.

NAMES OF AUTHORS INCLUDED IN THE LIST

The main list fundamentally follows the lay-out and style of Brummitt & Powell's book. The names of authors are quoted in roman type in the left hand column and the standard forms in boldface type in the right hand column.

It gives the names of all authors (known to me) who have validly established new taxa, new names, new combinations or other kind of *novitates* in the group of Pteridophyta. The majority of them are the same included in the list I compiled for the Working Party in 1991 and are now listed in *A.P.N.* with the code P (Pteridophyta). However, the wanting data concerning many of these authors, particularly their forenames and the dates of birth and death, has been in great part ascertained. Moreover the list has been enriched with many other names of authors who have established *novitates* after the publication in 1992 of Brummitt & Powell's book. Numerous other names of authors have also been added to the list.

The first batch of them consists of the authors of names established in periodicals, chiefly Chinese, which I had not been able to consult before the publication of *A.P.N.* and the authors of infraspecific taxa of ferns and fern allies which, not being listed in the *Index Filicum* and Supplements 1-4, had escaped my previous researches.

In addition to them I have included also the names of all authors of a composite paper even if only one of the co-authors is the real author of the *novitates* proposed there, as well as when in a composite paper the *novitates* established there are ascribed to an author who is not one of the co-authors of the paper. In the same way I have listed the names of the editor(s) of Floras and other books in which the *novitates* are ascribed to an author who is not a co-author of the book. However, I have declined to list the names of all co-authors of the big Floras, like *Flora Europaea* or *Flora of North America*, published by an Editorial Committee of many members.

To the list I have also added (duly marked) a certain number of authors who have proposed only names of ferns and fern allies not validly published. First among them are some pre-Linnaean botanists who were quoted by Linnaeus in his *Species Plantarum* (1753) and *Genera Plantarum* (1754) as authors of names or phrase-names of plants now regarded as belonging to the Pteridophyta. I have thought it convenient to list these authors because some names of their plants were taken up by Linnaeus as generic names or, duly reduced to the binomial nomenclature, as specific names, and because their specimens or drawings are regarded as the types or selected as the lectotypes of some Linnaean species. These names are printed in *italic* type.

Besides them I have also listed the names of other authors, mainly of recent times, who have proposed only new names which are invalid but are susceptible to be validated in the near future. This has been done since these authors' names will be included in the Sixth Supplement of the *Index Filicum* prepared by R.J.Johns at Kew. The names of these authors are marked in the list by an asterisk (✳) placed after their biographical data.

CITATION AND SPELLING OF AUTHORS' NAMES

In the main list, the citations of the names of authors are given in the main entry in the same way as in *A.P.N.*. Their surnames are followed by a comma, by the forename(s) and, in brackets, by the date(s) of birth and, when applicable, of death. Alternative names or spellings, or other variants of the surname or compound names cited only in part, are listed with cross references to the main entry. However, they are not quoted in brackets as in *A.P.N.*, since they are easy to recognize, the biographical data being given in the main entry only and never in the cross reference. Moreover, in the latter the word 'see' is replaced by '*vide*' (in italic type) and for the sake of brevity only the initials of the forenames are given there. For example:

Accepted name (main entry):
Palisot de Beauvois, Ambroise Marie François Joseph (1752-1820)   **P.Beauv.**

Alternative name (cross reference to main entry):
Beauvois Ambroise Marie François Joseph Palisot de
   vide Palisot de Beauvois, A.M.F.J.   **P.Beauv.**

As regards the names of women who have established new taxa or other kinds of *novitates*, they are listed in different ways. Those women who have published under their maiden names, even if married, are listed with their maiden names. Likewise, those women who have published under their married names are listed with their married names, but their maiden names, when known, are given in brackets preceded by the word 'née'. Women who have described some *novitates* both under their maiden and married names are listed with both of them. In this case, the maiden name is followed by the indication, in brackets, of the married name preceded by the word 'later', while the married name is followed by the indication, in brackets, of the maiden name preceded by the words 'née' [e.g. Das, Anjali (later Biswas, Anjali) and Biswas, Anjali (née Das, Anjali)]. In this case, according to principle no.11 established by the Working Party, different standard forms are given for the same person (e.g. A.Das and A.Biswas).

American women, when married, take the surname of their husband but keep their maiden surname (often quoted only by the initial) cited after the forename(s) [e.g. Tryon, Alice F. or Tryon, Alice Faber; Faber being her maiden surname)].

Some authors in their publications adopted only one (rarely two) of their official forenames, and some others added the translation in Latin or in other languages of their official forenames. Such names are included in the list in brackets after the names adopted in the publications [e.g. Ascherson, Paul (Friedrich August); Engler, (Heinrich Gustav) Adolf; Presl, Carl (Karel, Carel, Carolus), Bořivoj (Bořiwog, Bořiwag)].

Problems concerning some particular names

The problems concerning what constitutes the operative surname or how it should be spelled in the language of certain peoples have been taken into consideration in the introduction (pp.6-9) of Brummitt & Powell's book. These problems concern also the authors of ferns and fern allies, and the following notes are the result of some research I have carried out on their names, particularly the Chinese ones.

Cyrillic names

The romanization of Cyrillic words, i.e. the transliteration of Cyrillic words into Roman words, is a difficult problem since several different systems have been employed for the transliteration of Cyrillic names, and moreover the same Cyrillic letter can be transliterated into one Roman letter in one country and into a different Roman letter in another one (see Paclt, 1953). In accordance with *A.P.N.* I have preferred to include in the list a name of a Cyrillic-writing author with the same romanisation of his/her name which has been adopted in his/her publications. However, this is not easy to ascertain, since some authors have modified the transliteration of their name during their lifetime, and different romanisations of the name of an author have been used by other authors. Consequently, I have preferred to list the Cyrillic names of authors with the same spelling as that adopted in *A.P.N.* and I have quoted its various transliterations either in brackets after it, if the difference between them is slight [e.g. Tzvelev (Tzvelov, Tsvelev, Tsvelov)], or with a cross reference to it if they are rather different.

Chinese names

Most Chinese names are composed of a family name commonly of one single syllable placed first, followed by the forenames usually consisting of two single syllables. Very few family names have two syllables, and only a small percentage of forenames have a single syllable.

Chinese have only a few hundred family names, but thousands of forenames. According to Xu & Nicolson (1992) about 40% of Chinese have one of the following eleven family names: Chen, He, Hu, Li, Liu, Wang, Wu, Xu, Zhang, Zhao and Zhu. Brummitt & Powell cite 102 authors of plant names whose surname is Wang.

The romanisation of Chinese names, i.e. the conversion of Chinese ideograms into Roman alphabetic letters, is rather complicated. In fact Chinese names have been romanised in different ways, by different peoples and in different times. Moreover the pronunciation varies greatly in the different provinces. Consequently the romanisation can give rise to more than one spelling, sometimes to two or even three different spellings which are not easily recognizable as the name of a single person particularly when the difference concerns the initials of the surname (e.g. Chu, Wei-Ming and Zhu, Wei-Ming; Chang, Jin-Lun and Zhang, Jin-Lun; Xia, Chun and Hsia, Qun).

Various Chinese authors in their lifetime and sometimes in the same paper romanised their names by different spellings. For instance, the famous Chinese pteridologist mostly quoted his own name as Ching Ren-chang, but in some publications he adopted the variants Ching Renchang, Ching Ren-Chang, Qin Ren-chang and Chin Jen-chang. The name of another outstanding Chinese pteridologist Shing Kung-hsia has been romanised also as Shing Gong-hsia, Shing Hung-Hsia, Hsing Kung-hsia and Xing Gong-xia.

The romanisation of Chinese ideograms is mainly made according to two different conventions: the traditional English Wade-Giles and the modern Chinese phonetic alphabet system, better known as Pin-yin convention.

The Wade-Giles convention dates back to about the half of the last century, when in Europe, particularly in France, Germany and Great Britain, after the first attempts of the Jesuits, systematic research was undertaken to solve the problem of romanisation of Chinese ideograms. Among them the only one which passed over the frontiers of its own country and became later internationally accepted, was the system proposed by the English sinologist Sir Thomas Francis Wade in his book of 1867, revised in the second edition of 1886. There he introduced a phonetic romanisation of Chinese characters using apostrophes to denote the initial and final aspirated consonants.

In 1892, in the first edition of his *Chinese-English Dictionary* and in the second revised edition of 1912, H.A. Giles adopted a system, of romanisation which scarcely differs from that used by Wade. Likewise he employed apostrophes and accents, intending to render more accurately the sounds of the various letters of the Chinese alphabet. Giles adopted this system in his *Chinese Biographical Dictionary* published in 1898. There the authors' names were given with the surname followed by the two syllables of the forenames, joined by a hyphen, the first capitalized, the second in lower case (e.g. Chang Liang-chi;

Li Ts'un-hsü; Li Tzǔ-ch'êng). The main merit of Giles was that he adopted a system similar to that of Wade in his *Dictionary*. It had a great editorial success, so that he much contributed to the diffusion of this system which later began to be known as the Wade-Giles convention.

Some modifications to the Wade system were proposed by W.E. Soothill in 1899, B.Karlgren in 1928, C.S. Gardner in 1930, R.H. Mathews in 1931and by other sinologists. Merits and defects of these modifications were taken into consideration by the outstanding sinologist Joseph Needham in the first volume of his monumental work *Science and Civilisation in China*, first published in 1954, and reprinted five times, the last in 1979. In this volume, with the intent to "adhere to Wade as closely as possible" (p. 26), and with the wish "to avoid the extensive use of apostrophes characteristic of Wade and Giles to indicate the aspirated consonantal prefixes of words" (p. 26), Needham proposed a modified system, the characteristics of which are shown in a table of the "Romanisation of Chinese sounds" (p.24-25) where he compared the system adopted in his book with those of Wade-Giles and Gardner. The main modification consists in the addition of an *h* to the aspirated consonantal initials, so that *Ch'* becames *Chh* and *p'* becames *ph* and so forth, as well as in the use of the accents and the diaeresis for the sounds *ê, ǔ* and *ü* in particular names (e.g. Chhen Tê-Yün).

In the Bibliography "B. Chinese and Japanese books and journal articles since + 1800" (pp.263-267) as well as throughout the volume, Needham quoted the name of authors , giving first the surname, followed by the two syllables of the forenames, both capitalized and joined by a hyphen (e.g. Chang Yin-Lin). It is worth mentioning that in this Bibliography the author's names are accompanied by the Chinese ideograms.

Needham continued to use his modified Wade-Giles (in the following pages quoted as the Needham system) and the same style of citation of authors' name in the subsequent volumes of his work, including the first part of volume six devoted to Botany. This was published in 1986 when the Pin-yin convention had already been accepted. He wisely added to it (pp.710-718) two "Romanisation conversion tables" compiled by Robin Brilliant: one from Pin-yin to the modified Wade-Giles, and another from the modified Wade-Giles to Pin-yin.

The introduction of the Chinese phonetic alphabet system represented an improvement of the romanisation of Chinese ideograms in comparison with the Wade-Giles system, since it corresponds more closely to the way they sound to most western ears.

The Pin-yin convention was accepted in China as the official system of romanisation of Chinese language in 1958 (11 Feb.), and in 1971 an international committee was appointed for the preparation of a Pin-yin English Dictionary which was published in 1979 by Wu Jingrong, acting as editor-in-chief of the committee. It also contains a conversion table of "Chinese Phonetic Alphabet and Wade System" (pp.957-959).

In the romanisation of Chinese sounds, the Pin-yin differs greatly from the Wade-Giles system also because it makes no use of apostrophes and accents. Consequently, the romanisation of the ideograms according to the Pin-yin is less complicated than the Wade-Giles convention. This explains why it always tends to be more adopted throughout the world in humanistic literature.

The Pin-yin differs also in the citation of the authors' names. According to it the surname is followed by the two syllables of the forenames which are merged into a single word with only the first one capitalized (e.g. Ching Renchang). However, Xu in 1990 has proposed to capitalize also the second syllable (e.g. Xu ZhaoRan), in order to avoid possible ambiguities about the identity of some authors.

In plant taxonomic literature the romanisation of Chinese ideograms is almost exclusively made according the Wade-Giles or the Pin-yin conventions. Between them the number of authors who have selected the Wade-Giles system to quote their names in publications is by far greater than that of those who have adopted the Pin-yin system. Likewise, the majority of taxonomic bibliographies, indexes of herbaria or of plant collectors, even those published in recent times, give the names of authors according to Wade-Giles system.

In the selection of the system of romanisation of the names of Chinese authors, considering that the aim of this book is the publication of a list of names of authors who have established *novitates* in the pteridophytes, my preference has been given to the Wade-Giles convention, though I am aware that the use of the Pin-yin system will increase in the future.

In the previous pages I have briefly given some information on the main characteristics of the Wade-Giles system and its modifications particularly by Needham. Thus below I limit myself to dealing with the system adopted for the citations of names of Chinese authors.

The Wade-Giles system and its modifications have been greatly simplified and all the apostrophes, accents and additions of particular letters (e.g. the *h*) to express the sound of some aspirated consonants have been eliminated. In some recent publications, as well as in Brummitt & Powell's *Authors of Plant Names*, also the hyphen between the two syllables of the forenames has been omitted.

These simplifications are much contested by sinologists who consider them a mistake. I agree with them that they can give origin to ambiguity, but what is important in a work like this is to avoid confusion about the identity of the authors. This can be achieved by reference to bibliographies, biographies or other books in which the citation of authors' names is accompanied by the corresponding ideograms, and better still when it is provided with the indication of biographical data. With this information any doubt about the identity of authors is removed.

Two books published in 1993 allow us, at least in part, to gain this end: the *Bibliography of Chinese Systematic Botany (1949-1990)* by Chen Sing-chi *et al.* and the *Index Herbariorum Sinicorum* by Fu Li-Kuo *et al.*. In both of them the romanised Chinese names are accompanied by the corresponding ideograms.

In the citation of the romanised names according to the classical Wade-Giles system, the surname is followed by the two forenames which are given as two independent words joined by a hyphen, the first of them written with a capital initial and the second with a small letter (e. g. Ching Ren-chang). This system is adopted in the *Bibliography* by S.C. Chen (1993) cited above.

In the citations of the romanised names according to the Needham system the surname is followed by the two syllables of the forenames which are treated as two independent words joined by a hyphen, both written with capital initials (e.g. Ching Ren-Chang). This system is adopted in the *Index* by L.K. Fu *et al.* (1993), to which I have made reference.

During my research on Chinese botanists, hunting for authors of scientific names of ferns and fern allies, I have ascertained that the romanisation of their names is mostly made according to the classical Wade-Giles system. However, in recent times a clear tendency has emerged to prefer the Needham system. Between them I have decided to adopt the last one. I have been induced to take this decision for various reasons: in first place, because this system is used in the above mentioned *Index* by L.K. Fu *et al.*, which offers a greater certainty about the identity of the authors, since their biographical data are given in addition to the romanised Chinese names and the corresponding ideograms. Furthermore because in the book many names of other botanists are given in the list of collectors, and this system is nearer to that adopted in *A.P.N.*.

Even if I have followed the system adopted by Fu *et al.* in my research I have carefully examined the book of Chen *et al.* and I have drawn much advantage from the great wealth of information assembled there.

Xu & Nicolson (1992) have recommended that no Chinese names be abbreviated, but one of the aims of *A.P.N.* and the present book is to establish the standard forms of authors of plant names, and while the Chinese surname is always here given in full, abbreviation of the forenames cannot be avoided.

Jin S.Y. in his paper *Bibliography of new taxa (Cormophytes) type specimens in the herbaria of China (I)* (1988), and Wu S.H. & Ching R.C. in their book *Fern families and genera of China* (1991) quote in full the surname of the authors but abbreviate their forenames. Although the former adopts the Needham system and the latter the classical Wade-Giles, both of them give the abbreviation of the authors' names by the initials of the two syllables of the forenames not hyphenated followed by the surname (e.g. R.C. Ching); a kind of abbreviation which quite agrees with that of the standard forms recommended in *A.P.N.* and in this book.

Compound names of Spanish, Portuguese and other origin

One of the best sources of information on the names of Spanish and Portuguese authors is the Appendix to the first four volumes of *Flora Iberica* (Castroviejo *et al.* 1986-1993) where rich information and standard forms of these names are given.

The peoples of Spanish origin use two surnames: the first comes from the first surname of the father, while the second comes from the first surname of the mother [e.g. *forename:* Paloma — *surnames:* Cubas (father's) Domínguez (mother's)].

The peoples of Portuguese origin equally use two surnames, but the first comes from the first surname of the mother and the second from the first surname of the father [e.g. *forenames*: João Manuel Antonio — *surnames*: Paes do Amaral (mother's) Franco (father's)].

The two peoples use both the surnames for official and legal purposes, and consider the father's name as the most important and the effective surname of the family. As regards the mother's surname in scientific publications some authors wish to keep it together with that of the father, others prefer to omit it and to use the father's surname only. Consequently, in many cases it is uncertain how a name of an author has to be listed, also because some authors in their publications have adopted both the surnames or that of the father only.

My research in the original publications and by direct correspondence with some authors has solved some problems about certain names, but I doubt that I have given satisfactory and correct citations of the names of other authors, in spite of the high number of cross reference I have given for Spanish and Portuguese authors.

Another problem to be mentioned here concerns the citation of the complicated and long compound names of certain French authors of the 18th-19th centuries, such as Palisot de Beauvois, Du Petit-Thouars, Lamarck, etc.. The problem is not limited to the sequence of the various surnames and by the prefix 'de' which accompany them, but sometimes also by the definition of the forenames. The surnames of these authors are quoted in literature in the most diversified manner. I have devoted my attention to this problem, particularly in various research I have carried on in some libreries of Paris and I consider that at least some citations of these names, certainly not all of them, given in my list are reliable. These names have required several cross references to the main entry.

Names provided with prefixes

Divergent opinions were expressed within the Working Party over whether the prefixes 'van', 'van der', 'van den' and 'de' should be kept in standard forms of author's names or not, and whether, when included, they should be capitalized or spelled with lower case initials. The opinions were discordant also because the peoples using these prefixes belong to different countries.

All efforts to gain general consensus were unsuccessful, but at least from the various conflicting views emerged the possibility to adopt some principles, not binding and differing from each other, for the use of these prefixes in the standard forms of the Dutch, Belgian and South African authors. These principles have been listed by Brummitt & Powell in the introduction of *A.P.N.* (p.8), and are reported here as follows.

Dutch — 'van' and 'van der' to be eliminated while 'de' be retained and all spelled with lower case initial.

Belgian — 'Van', 'Van der' and 'De' to be retained, all spelled with upper case initial.

South African — 'van', 'van der' and 'de' to be retained, spelled with a capital initial if no forename initial precedes them, but with lower case initial if forename initials are included.

As regards the remaining names of authors of other nationalities, although the prefixes were mostly retained, no principles were established and many cases were taken into consideration separately. No general rule was envisaged about their capitalization.

As concerns names provided with prefixes, examination of the original publications and some research by correspondence allowed me to ascertain the spelling to use for the surnames of certain authors. Some of them turned out different from those cited in *A.P.N.* and I have reversed the main entry and the cross reference. However, the standard forms recommended in *A.P.N.* have been always retained.

In spite of all efforts to establish some general principles for the spelling of names with prefixes, several of them still remain uncertain. Very wisely in *A.P.N.* the prefixes are mostly omitted in their standard forms in order to avoid future changes in them. A great number of cross references are given in order to trace them.

As regards the prefix 'von' of German-speaking authors it was consistently omitted in the recommended standard forms, but it was quoted in the part of the entries where surnames and forenames are given in full.

STANDARD FORMS

As regards the standard forms I have rigidly followed the above-mentioned principles established by the Working Party, both for the names listed in Brummitt & Powell's book for the authors of names in Pteridophytes (marked there by P) and for the authors added to the list after the publication of *A.P.N.*. However, I have corrected some forenames which turned out to be erroneous or not in accordance with the above-mentioned principles (e.g. Barnola, not Joaquin; d'Almeida, not J.F.R.Almeida; Enys, not Emys; Koch-Grünb., not G.C.T.Koch; Port.-Led., not Port.). All these standard forms were corrected in the database of the Royal Botanic Gardens, Kew.

SELECTED BIBLIOGRAPHY

This list contains the references to the publications quoted in the Preface and in the Introduction, as well as the bibliographical citations of those books and papers selected among the hundreds examined during my research, which proved to be particularly helpful to tracing the author's names of ferns and fern allies and ascertaining the forenames and biographical data of many authors unknown up to now.

Adanson, M. — *Familles des Plantes*. Paris, 1763.
Barnhart, J.H. — *Biographical notes upon botanists*. vol. 1-3. Boston, 1905.
Britten, J. & Boulger, G.S. — *A biographical index of deceased British and Irish Botanists*. ed.2. London, 1931.
Broun, M. — *Index to North American Ferns*. Orleans, Mass., 1938.

Brummitt, R.K. & Powell, C.E. — *Authors of Plant Names*. (Roy. Bot. Gard., Kew). Whitstable, 1992. (Quoted as *A.P.N.*).

Burdet, H.M. — *Auxilium ad botanicorum graphicem*. Genève, 1979.

Burdet, H.M. & collab. — *Ouvrages Botaniques Anciens*. Genève, 1985.

Castroviejo, S. *et al.* — *Flora Iberica*. vol. 1-4. Madrid, 1986-1993.

Chen, Sing-Chi, Li, Jao-Lan *et al.* — *Bibliography of Chinese systematic Botany*. Guangzhou, 1993.

Ching, Ren-Chang, Shing, Kung-Hsia *et al.* — *Flora Reipublicae Popularis Sinicae*. Tomus 2 & 3(1). Pteridophyta. Beijing, 1959-1990.

Christensen, C. — *Index Filicum*. Copenhagen, 1905-1906.

Christensen, C. — *Index Filicum. Supplementum 1906-1912*. Copenhagen, 1913.

Christensen, C. — *Index Filicum. Supplementum preliminaire pour les années 1913, 1914, 1915, 1916*. Copenhagen, 1917.

Christensen, C. — *Index Filicum. Supplementum tertium pro annis 1917-1933*. Copenhagen, 1934.

Clusius, C. — *Rariorum aliquot stirpium per Hispanias observatarum historia*. Antwerpiae (Anvers), 1576.

Crosby, M.R., Magill, R.E. & Bauer, C.R. — *Index of Mosses. 1963-1989*. Monogr. Syst. Bot. Missouri Bot. Gard. vol. 42. 1992.

Czerepanov, S.K. — *Plantae Vasculares Rossicae et civitatum collimitanearum (in limitis URSS olim)*. S.Petropolis, 1995.

Dale, E.J. — *Literature on the history of botany and botanic gardens 1730-1840. A Bibliography*. Huntia 6: 1-121. 1985.

Davies, R.A. & Lloyd, K.M. — *Kew Index for 1986 to 1989*. Oxford, 1987-1989.

Davy de Virville, A. — *Histoire de la Botanique en France*. Paris, 1954.

Desmond, R. — *Dictionary of British and Irish Botanists and Horticulturists including Plant Collectors and Botanical Artists*. London, 1977.

Desmond, R. & collab. (Elwood, C.) — *Dictionary of British and Irish Botanists and Horticulturists including Plant Collectors, Flower Painters and Garden Designers*. London, 1994.

Dörfler, J. — *Botaniker-Adressbuch*. Wien, 1896; ed. 2. 1909.

Durante, C. — *Herbario Nuovo*. Roma, 1585.

Encke, F., Buchheim, G. & Seybold, S. — *Zander Handwörterbuck der Pflanzennamen*. ed.15. Stuttgart, 1994.

Fiori, A. — *Flora Italica Cryptogama. Pars V: Pteridophyta*. Firenze, 1943.

Fu, Li-Kuo, Zhang, Xian-Chun *et al.* — *Index Herbariorum Sinicorum*. Beijing, 1993.

Gardner, C.S. — *A modern system for the Romanisation of Chinese*. Harvard, 1930.

Giles, H.A. — *A Chinese Biographical Dictionary*. 2 vols. (Kelly & Walsh) Shanghai, & (Quaritch) London, 1898.

Giles, H.A. — *A Chinese-English Dictionary*. ed.2. 4. Part I & Part II (3 vols.). (Kelly & Walsh), Shanghai, & (Quaritch) London, (1909-1910, 1911) 1912.

Gould, S.W.& Noyce, D.C. — *International Plant Index. vol.1. Family names of the plant kingdom - vol. 2. Authors of plant genera*. New York, 1962, 1965.

Grimes, J.W. & Parris, B.S. — *Index Thelypteridaceae*. (Roy. Bot. Gard., Kew). Whitstable, 1986.

Heller, J.L. — *Index auctorum et librorum a Linnaeo Species Plantarum, 1753 citatorum*. In Heller, J.L. & Stearn, W.T., *An appendix to the Species Plantarum of Carl Linnaeus*. In Linnaeus, C., *Species Plantarum. A facsimile of the first edition 1753*. vol. 2. London (Ray Society), 1959.

Holman, J.H. & Jermy, A.C. — *An international directory of Pteridologists*. London, 1973 (Mimeogr.).

Holmgren, P.K., Holmgren, N.H. & Barnett, L.C. — *Index Herbariorum. Part I: The herbaria of the World*. ed. 8. New York, 1990.

Huitième Congrès International de Botanique Paris 1954. *Actes du Congrès et dernières comunications reçues. Résolution relative à l' «Index Filicum» présentée par la Section 4. Taxinomie, Systématique et Philogenie*. p.115, Paris, 1959.

International Association for Plant Taxonomy — *Index Filicum (resolution)*. Taxon 4(3): 70. 1955.

International Association for Plant Taxonomy — *Index Filicum Committee*. Taxon 4(7): 178-179. 1955.

Jackson, B.D. — *Guide to the Literature of Botany*. London, 1881.

Jarrett, F.M. & collab. (Bence, T.A., Grimes, J.W., Parris, B.S. & Pinner, J.L.M.) — *Index Filicum. Supplementum Quintum pro annis 1961-1975*. Oxford, 1985.

Jermy, A.C. (edit.) — *The I.A.P. Pteridophyte Bibliography 1982/83*. London, 1985.

Jin, Shu-Ying — *Bibliography of new taxa (Cormophytes) and type specimens in the Herbaria of China (I)*. Bull. Bot. Res. Harbin. Addit.1-119. 1988.

Johns, R.J. — *Index Filicum. Supplementum Sextum pro annis 1976-1990*. (Roy. Bot. Gard., Kew) (in press).

Karlgren, B. — *The Romanisation of Chinese*. (China Soc.). London, 1928.

Kew Record of Taxonomic Literature (The) relating to the vascular plants [for 1971 to 1995(3)]. (Roy. Bot. Gard., Kew). London, 1974-1995.

Kiger, R.W., Jacobsen, T.D. & Libby, R.M. — *International Register of Specialists and Current Research in Plant Systematics*. Pittsburgh, 1981.

Knobloch, I.W. — *Pteridophyte Hybrids*. Publ. Mus. Michigan State Univ. Biol. Ser. 5(4): 273-352. 1976.

Knobloch, I.W. — *Pteridophyte hybrids and their derivatives*. (Michigan State Univ.). East Lansing, 1996.

Komarov, V.L. *et al.* — *Flora U.R.S.S. (Flora Unionis Rerumpublicarum Socialisticarum Sovieticarum)*. vol.1-30. Leningrad, 1934-1964.

Kurata, S. & Nakaike, T. (edits.) — *Illustrations of Pteridophytes of Japan*. vol.1-7. Tokyo, 1979-1994.

Lanjou, J., Stafleu, F.A., Chaudheri, M.N., De Wal, C.M. & Vegter, I.H. — *Index Herbariorum. Part II. Collectors*. No.1-7 (A-Z). Regnum Veg. vol. 2, 9, 86, 93, 109, 114, 117. Utrecht, 1954-1988.

Laundon, J.R. — *Deceased lichenologists: their abbreviations and herbaria*. Lichenologist 11(1): 1-26. 1979.

Lowe, E.J. — *Our native Ferns.* vol.1-2. London, 1867, 1869.

Mabberley, D.J. — *The plant-book.* Cambridge, 1990.

Margadant, W.D. — *Early Bryological Literature.* Utrecht, 1968.

Mathews, R.H. — *Chinese-English Dictionary.* (China Inland Mission). Shanghai, 1931.

Meikle, R.D. — *Draft Index of Author Abbreviations compiled at The Herbarium Royal Botanic Gardens, Kew.* Basildon, 1980.

Merrill, E.D. & Walker, E.H. — *A bibliography of Eastern Asiatic botany.* Jamaica Plain Mass. USA, 1938. Supplement I (by Walker E.H.). Washington, 1960.

Morin, N.R. (Convening Editor) & Editorial Committee — *Flora of North America, North of Mexico.* vol.1-2. New York, 1993.

Morison, R. — *Plantarum Historiae Universalis.* Vol. 3. Oxonia 1699.

Nakaike, T. — *Enumeratio Pteridophytarum Japonicarum. Filicales.* Tokyo, 1975.

Needham, J. — *Science and Civilisation in China.* vol. I. *Introductory Orientations.* (First published in 1954). Reprint. ed. (Cambridge Univ. Press). Cambridge 1979.

Needham, J. & collab. (Lu, Gwei-Djem & Huang, Hsing-Tsung) — *Science and Civilisation in China.* vol.6. *Biology and Biological Technology.* Part I: *Botany.* (Cambridge Univ. Press). Cambridge, 1986.

Nitsche, E.M., Taylor, W.C., Moran, C.K.R. & Moran, R.C. (edits.) — *Annual Review of Pteridological Research.* vol. 1-7. 1987-1993.

Ohwi, J. — *Flora of Japan.* Washington, 1965.

Paclt, J. — *Transliteration of Cyrillic for use in botanical nomenclature.* Taxon 2(7): 159-166. 1953.

Pichi Sermolli, R.E.G. & collab. (Ballard, F., Holltum, R.E., Itô, H., Jarrett, F.M., Jermy, A.C., Schelpe, E.A.C.L.E., Tardieu-Blot, M.-L. & Tryon, R.M.) — *Index Filicum. Supplementum quartum pro annis 1934-1960.* Regnum Veg. vol.37. Utrecht, 1965.

Pichi Sermolli, R.E.G. — *A provisional catalogue of the family names of living Pteridophytes.* Webbia 25(1): 219-297. 1970.

Pichi Sermolli, R.E.G. — *Historical review of the higher classification of the Filicopsida.* In Jermy, A.C., Crabbe, J.A. & Thomas, B.A. (edits.) —*The phylogeny and classification of the ferns.* Bot. Journ. Linn. Soc. London 67(Suppl.1): 11-40. tav.1-19. 1973.

Pichi Sermolli, R.E.G. — *Tentamen Pteridophytorum genera in taxonomicum ordinem redigendi.* Webbia 31(2): 313-512. 1977.

Pritzel, G.A. — *Thesaurus Literature Botanicae.* ed.2. Lipsiae, 1872-1877.

Quinby, J. & Stevenson, A. — *Catalogue of Botanical Books in the collection of Rachel McMasters Miller Hunt. Catalogue of the Hunt Botanical Collection.* Pittsburgh, vol. 1 (J. Quinby), 1958; vol. 2(1-2) (A. Stevenson), 1961.

Reed, C.F. — *The phylogeny and ontogeny of the Pteropsida. I. Schizaeales.* Bol. Soc. Brot. ser. 2. 21: 71-197. 1948.

Reed, C.F. — *Index Isoëtales.* Bol. Soc. Brot. ser. 2. 27: 5-72. 1953.

Reed, C.F. — *Index Marsileata et Salviniata*. Bol. Soc. Brot. ser. 2. 28: 5-61. 1954.

Reed, C.F. — *Index Marsileata et Salviniata. Supplementum*. Bol. Soc. Brot. ser. 2. 39: 259-302. 1965.

Reed, C.F. — *Index Psilotales*. Bol. Soc. Brot. ser. 2. 40: 71-96. 1966.

Reed, C.F. — *Index Selaginellarum*. Mem. Soc. Brot. 18: 1-287. 1966.

Reed, C.F. — *Index Thelypteridis*. Phytologia 17(4): 249-328. 1968.

Reed, C.F. — *Index Equisetophyta. Part II: Extantes. Index Equisetorum*. Baltimore, 1971.

Saccardo, P.A. — *La Botanica in Italia*. Venezia (Mem. Ist. Veneto Sci. Lett. Arti 25(4): 1-236). 1895.

Saccardo, P.A. — *La Botanica in Italia. Parte seconda*. Venezia (Mem. Ist. Veneto Sci. Lett. Arti 26(6): 1-172). 1901.

Salvo, A.E., Asensi, A. & Rivas-Martinez, S. — *Bibliografia Pteridologica de España Portugal (Continente e islas) 1802-1980*. Trab. y Monograf. Dept. Bot. Malaga 2: 59-104. 1981.

Săvulescu, T., Nyárády, E.J. & Pop, E. (edits.) — *Flora Reipublicae Socialisticae România*. vol.1-13. Bucarest, 1952-1976.

Sayre, G. — *Authors of names of Bryophytes and the present location of their herbaria*. Bryologists 80(3): 502-521. 1977.

Small, J.K. — *Ferns of the Southeastern States*. Lancaster, 1938.

Soothill, W.E. — *The student's four thousand characters and general pocket dictionary*. (Presbiterian Mission Press). Shanghai, 1899.

Stafleu, F.A. & Cowan, R.S. — *Taxonomic Literature*. vol. 1-7. Regnum Veg. vol. 94, 98, 105, 110, 112, 115, 116. Utrecht, 1976-1988. (Quoted as *TL-2*).

Stafleu, A.F. & Mennega, E.A. — *Taxonomic Literature. Supplements I. II. III*. Regnum Veg. vol. 125, 130, 132. Königstein, 1992-1995.

Stewart, R.R. — *An annotated catalogue of the vascular plants of West Pakistan and Kashmir*. In Nasir, E. & Ali, S.I. (edits.), *Flora of West Pakistan*. Karachi, 1972.

Townsend, C.C., Guest, E. & Al-Rawi, A. — *Flora of Iraq*. vol. 1-2. Baghdad, 1966.

Tutin, T.G. *et al.* (edits.) — *Flora Europaea*. vol. 1-5. Cambridge, 1964-1980.

Tutin, T.G. *et al.* (edits.) — *Flora Europaea*. ed. 2. vol.1. Cambridge, 1993.

van Steenis-Kruseman, M.J. — *Malaysian Plant Collectors and Collection, being a Cyclopaedia of Botanical exploration in Malaysia*. Flora Malesiana, ser. I, vol. 1. 1950.

van Steenis-Kruseman, M.J. — *Malaysian Plant Collectors & Collections. Supplement I*. ser. I. vol. 5(4): pp.CCXXXVII-CCCXLI. 1958.

Verma, S.C., Khullar, S.P. *et al.* — *Pteridology in India. A bibliography*. Dehra Dun, 1987.

Wade, T.F. — *A progressive course designed to assist the student of colloquial Chinese as spoken in the Capital and Metropolitan department*. First ed., London, 1867; ed. 2. 3 vols. Shanghai, 1886.

Wagenitz, G. — *Index Collectorum principalium Herbarii Gottingensis*. Göttingen, 1982.

Walker, E.H. — *Flora of Okinawa and the Southern Ryukyu Islands.* Washington, 1976.

Weatherby, C.A. — *List of varieties and forms of the Ferns of Eastern North America*. Amer. Fern Journ. 25: 45-51; 95-100. 1935 — 26: 11-16; 60-69; 94-99; 130-136. 1936 — 27: 20-24; 51-56. 1937.

White, R.A., Lloyd, R.M., Cousens, M.I., Haufler, C.H., Brook, R.E., Taylor, W.C. & Nitsche, E.M. — *Bibliography of American Pteridology*. vol.1-12. 1976-1987.

Wittrock, V.B. — *Catalogus illustratus Iconothecae Botanicae Horti Bergiani Stockholmiensis. Notulis biographicis adjectis*. Pars I: Acta Horti Berg. 3(2): 1-198. tav.1-37. 1903. — Pars II: 3(3): I-XCIII + 1-245. tav.150. 1905.

Wu, Jingrong (Editor-in-chief & Editorial Committee) — *The Chinese-English Dictionary*. (Commercial Press Ltd.) Hong Kong & (Pitman Publ. Ltd.). London, 1979.

Wu, Shiew-Hung & Ching, Ren-Chang — *Fern families and genera of China*. Beijing, 1991.

Xu, Zhao-Ran — *A discussion on the romanization of Chinese personal names*. Guihaia 10(1): 87-91. 1990.

Xu, Zhao-Ran & Nicolson, D.H. — *Don't abbreviate Chinese names*. Taxon 41(3): 499-504. 1992.

SOME STATISTICAL DATA

A.P.N. = Brummitt R.K. & Powell C.E. - Authors of Plant Names, 1992.

Novitates = New taxa at all ranks, new names, new combinations, new status, and new hybrids.

P = Code Group in *A.P.N.* for Pteridophyta.

vide = see (in cross references).

Pichi Sermolli R.E.G.
Index Filicum. Supplementum quartum, 1965: pp.351-361

Authors of *novitates* in ferns (genera, infrageneric taxa and species
  of ferns) **802**

Pichi Sermolli R.E.G. in Brummitt R.K. & Powell C.E.
Authors of Plant Names, 1992

Authors of *novitates* listed with P in *A.P.N.* (ferns and fern allies) **1810**
Variants of names, cross references of them to the main entries
  and pre-Linnaean authors **52**

Total **1862**

Pichi Sermolli R.E.G. (this book, 1996)

Authors of *novitates* (ferns and fern allies at all ranks including infraspecific names and hybrids)	**2211**
Variants of names, cross references of them to the main entries (*vide*), pre-Linnaean names and names quote in *A.P.N.* by mistake	**576**
Total of the entries	**2787**

Comparison of the data in the present book and in A. P. N.

Authors of *novitates* in Pteridophyta in this book	**2211**
Authors of *novitates* in *A.P.N.* (including 6 names listed by mistake)	**1810**
Authors of *novitates* in Pteridophyta recorded for the first time in this book	**401**

SUBDIVISION OF THESE NEW AUTHORS

1) Authors already recorded in *A.P.N.* but not for P	**167**
2) Authors of *novitates* in Pteridophyta not yet recorded in *A.P.N.*	**162**
3) Pre-Linnaean authors recorded in this book	**45**
4) Authors of other invalid names recorded in this book	**27**
Total	**401**

ADDITIONAL DATA

Biographical data ascertained for authors quoted in *A.P.N.* with "floruit" only	**231**
Forenames in full ascertained for authors given in *A.P.N.* with the initials only	**189**

APPENDIX

During the compilation of the list of authors I realized that in the past the same abbreviation or the same surname was used for two different persons (e.g. Bert. for C.L.G.Bertero and A.Bertoloni; Gmel. for J.F.Gmelin and S.G.Gmelin; Koch for K.H.E.Koch and W.D.J.Koch; Scott for J.Scott and R.R.Scott) with the consequence that it was not always easy to establish which new name had to be ascribed to one author and which to the other.

I also ascertained that the abbreviations of some names were reduced to only two or three letters, sometimes without a vowel (e.g. B.S.P. used for N.L.Britton, E.E.Sterns & J.F.Poggenburg; Kl. for J.F.Klotzsch; PB. for

A.M.F.J.Palisot de Beauvois; v.d.B. for R.B. van den Bosch), and it was difficult to recognize the authors quoted by those abbreviations. Moreover, I realized that the name of an author had been given by different abbreviations. Finally, I found that several abbreviations of names adopted in *TL-2* actually had not been retained by the Working Party in spite of the recommendation to adopt them (see principle no.5).

These considerations induced me to prepare a list of these abbreviations with the name of the pertinent author and the corresponding standard forms accepted in *A.P.N.* and to publish it as an Appendix to this book.

ACKNOWLEDGEMENTS

This book is issued as one of the Kew Scientific Publications and I wish to express my most sincere gratitude to the Director (Prof. G.T.Prance) and the General Editor (Dr. J.M.Lock) of the Royal Botanic Gardens, Kew for this great honour.

I am also extremely grateful to my collaborator Prof. Maria Paola Bizzarri whose skilful and unselflish assistance in the bibliographical research, in the typesetting and in proof reading was particularly important for the publication of this book. I am also greatly indebted to Prof. Kung-Hsia Shing, as well as to Dr. Xian-Chung Zhang for their invaluable collaboration for establishing the biographical data and spellings of the names of several Chinese authors.

Particular thanks are due to the Directors and Staff of the Libraries of the Royal Botanic Gardens, Kew, of the National History Museum of London, of the Museum d'Histoire Naturelle de Paris, the Conservatoire et Jardin Botanique de Genève, and of the Biblioteca Botanica dell'Università di Firenze for the hospitality and facilities I have enjoyed during my research.

My most sincere thanks are addressed also to my friend Dick (Dr. R.K.) Brummitt for advice on various problems of spellings of authors' names, for information on several recent authors of names in Pteridophyta, for the revision of the preface and introduction of this book, and for the kind help received in many different ways during my visits to Kew. My thanks are extended also to another friend at Kew, Bob (Prof. R.J.) Johns, who provided me with a copy of his draft of the Sixth Supplement of the *Index Filicum* and of his work on the publications on Pteridophytes, 1975-1994, which proved very useful as a source of information on many recent authors of names of ferns and fern allies.

I am particularly indebted, to prof. Edoarda Masi (Milano), prof. Federico Masini (Roma) and dr. Giovanna Aleandri (Firenze) for the advice and assistance in the preparation of the account on the problems of Chinese names.

I am also very grateful to the following authors who have replied to my letters providing me with the forenames and biographical data of various authors of names in Pteridophytes: S.M. Almeida* (Bombay), F. Badré (Paris), P. Bamps (Meise), P. Berthet (Lyon), S.S. Bir* (Patiala), J. Braggins (Auckland), J.E. Burrows (Lyndenburg), S. Castroviejo (Madrid), H.T. Clifford (S.Brisban), D.S. Conant* (Lyndonville, USA), P. Cubas (Madrid), R.

Desmond* (Twickenham), R.D. Dixit (Allahabad), C. Favarger (Neuchâtel), R. Fernandes (Coimbra), J.M.A. Franco (Lisboa), H.P. Fuchs-Eckert* (Trin-Vigt), J. Gamisans (Marseille), C.B. Gena (Ajmer), B. Ghosh* (Calcutta), G.E. Giudice (La Plata), H.K. Goswami (Bhopal), J.L. Gradaille (Soller), R.J.F. Henderson (Indooroopilly), E. Hennipman (Bilthoven), P.H. Hovenkamp (Leiden), L.G. Hickok (Amherst), R.W. Hobdy (Wailuku), P.K. Holmgren (New York), J. Holub (Prühonice, Praha), C.R. Huxley (Oxford), Y.A. Ivanenko* (St.Petersburg), K. Iwatsuki* (Tokyo), C.E. Jarvis (London), P.M. Kirk (Egham), Ph. Küpfer (Neuchâtel), S. Kurita (Chiba), V.B. Kuvaev (Moscow), J.X. Li* (Shandong), D. Löve (San José, USA), Director of Institute of Genetics of University of Lund, P.V. Madhusoodanam (Kerala), D.A. Madulid (Manila), R.M. Masalles (Barcelona), J.F. Matthews (Charlotte), B.N. Mehrotra (Lucknow), F. Menegalle (Padova), J.T. Mickel (New York), I. Moreira (Lisboa), T. Nakaike* (Chiba), J.M. Nieto-Caldera (Malaga), Y.-C. Oh (Seoul), B. Øllgaard (Risskov), Y.P.S. Pangtey (Naini Tal), G. Panigrahi* (Calcutta), B.S. Parris (Kerikeri), M. Ponce (San Isidro), C.A. Raine (Manchester), R.R. Rao (Lucknow), H. Rasbach (Glottertal), K. Rasbach (Glottertal), J.F. Reed (New York), C.H. Rolleri (La Plata), J.A. Rosselló (Valencia), C. Sánchez Villaverde (Habana), K.H. Shing* (Beijing), P.S. Short (Victoria, Australia), P.K. Shukla (Allahabad), D.K. Singh (Dehra Dun), I.V. Sokolova (St.Petesburg), M.D. Tindale (Sydney), W.H. Wagner Jr.* (Ann Arbor), S.F. Wu (Shanghai), P.M. Zamora (Quezon City), X.C. Zhang* (Beijing). — Among them, I am particularly grateful to those marked with an asterisk (*) who have contributed to my research with the original data on numerous authors from China, Europe, India, Japan, North America and Russia, sometimes making difficult investigations to satisfy my request for information.

My thanks are also extended to many colleagues interested in pteridology, too numerous to mention individually, who have sent me the data needed on their names and year of birth.

Finally, I wish to express my gratitude to Mrs. Giuliana Ceccantini of the Dipartimento di Biologia Vegetale dell'Università di Firenze who has assisted Prof. P.Bizzarri in typesetting the list of authors.

EXPLANATIONS FOR THE USE OF BOOK

This book consists of a list of the authors of scientific names in Pterido-phyta, new taxa, new combinations, new names, new hybrids, etc., and an Appendix with the list of the abbreviations of the authors' names of the past and corresponding standard forms accepted in the *A.P.N.* and this book.

The following explanations represent a synthesis of the principles, rules and recommendations adopted in the compilation of the list. Details on these subjects, on the reasons for some choices, and on the solutions of some problems are given in the introduction to which I refer the reader for further information.

The main intention of these explanations is, in fact, to supply a sort of guide for a quick consultation of the List and the Appendix.

In the list the authors of names are given in Roman type and arranged in alphabetical order by their surnames (family names or last names) in the left hand column of the page. They are followed by a comma and by their full forenames (given names), when known, or by their initials when research to establish full forename was unsuccessful. Their dates of birth and/or death, as applicable, and as far as have been ascertained, are given in brackets after these data. The recommended standard forms of authors' names are given in boldface in the right hand column.

Alphabetisation is made by computer; accents ignored, upper and lower case are treated as the same; apostrophes and hyphens sorted before letters. Mc and St. abbreviations of surnames are not expanded to Mac and Saint. As regards the hyphen, the arrangement of the names in alphabetical order is important. In fact, for example, a hyphenated compound surname is treated as one word while the same compound surname not hyphenated is considered as two distinct surnames, with the consequence that they can be placed far from each other.

In the list I have also included some names of pre-Linnaean authors. Their number is limited to those who are the authors of names or phrase-names of plants now regarded as belonging to the Pteridophyta. They are listed in *italic* type.

I have also included in the list the names of certain authors, mostly alive, who have proposed, but not validly published, only new taxa or other *novitates* which are susceptible to being validated in the near future. They are marked in the list by an asterisk (✻) placed after their biographical data.

As mentioned in the introduction, the citation and spelling of some particular names have one or more variants or alternative surnames which for one reason or another differ from those which have been selected in *A.P.N.* and in this book as the accepted surnames. These variants or alternative names are given in the list either by their citation in brackets after the accepted surname, or by a cross reference to the accepted surname quoted in the second line of the cross references preceded by the word *vide* (in italic type). For example:

> *Accepted name (main entry):*
> Pérez Arbelàez, Enrique (1896-1972)   **Pérez Arbel.**
>
> *Alternative name (cross reference to main entry):*
> Arbelàez, Enrique Pérez
> *vide* Pérez Arbelàez, E.   **Pérez Arbel.**

Certain alternative names derived from the transliteration of Cyrillic names, can give rise to different spellings of the name of a single person; among them one is regarded as the accepted name. Several of these variants slightly differ slightly from each other and are mostly given in brackets after the accepted spelling of the surname.

Chinese names have been romanised by different systems; the best known of them are the Wade-Giles and Pin-yin conventions. However, following a tendency which has emerged in recent times, I have adopted the Wade-Giles system modified by Needham, in which the two syllables of the forenames are written with capital initials and are joined by a hyphen.

The use of one or another of the various systems adopted for the romanisation of Chinese letters in past and recent publications has given rise to many alternative names, some of them so different as to be difficult to recognize as the names of a single author. Consequently, cross references are largely employed for these variants of Chinese names.

Spanish and Portuguese authors use compound official family names consisting of the surnames of the father and the mother. In their publications some of them adopt both these official surnames, while others prefer to use only the surname of the father. Consequently, many cross references are needed in order to avoid ambiguities.

Certain French authors of the 18th-19th centuries have complicated and long compound surnames which have been quoted in a very diversified manner and two or three, even four distinct cross references for each of them are needed to clarify the diversity of these variants.

Authors' names provided with the prefixes such as 'de', 'van', and 'van der', represent a further difficult problem since the prefix before the surname is retained for some authors and not for others. As mentioned in the introduction different principles were adopted for Dutch, Belgian and South African names, while no principle was established for the use of prefixes for names of authors of other nationalities, although the prefixes were mostly retained. Many names were taken into consideration separately. Cross references to the accepted names were often used to explained where the prefixes were retained or omitted.

The German 'von' was generally eliminated in the standard forms, but often quoted in the entry where full surnames and forenames are given.

The biographical data are given always in brackets. They consist of the year of birth and/or, when applicable, of death. When both of them are known they are quoted one after the other separated by a hyphen. However, when neither date of birth nor of death is known, any one date when the author is known to have published a name is given after the abbreviation 'fl.' ('floruit'). The date of birth of a living author is followed by a hyphen and a blank space. One question mark means an unknown or uncertain date.

LIST OF AUTHORS OF SCIENTIFIC NAMES
IN PTERIDOPHYTA

Abbe, Ernst Cleveland (1905-   )   **Abbe**
Abbiatti, Delia (1918-   )   **Abbiatti**
Abeywickrama, Bartholomeusz Aristides (1920-   )   **Abeyw.**
Abrams, LeRoy (1874-1956)   **Abrams**
Acharius, Erik (1757-1819)   **Ach.**
Adams, Charles Dennis (1920-   )   **C.D.Adams**
Adanson, Michel (1727-1806)   **Adans.**
Adelbert, Albert George Ludwig (1914-1972)   **Adelb.**
Afzelius, Adam (1750-1837)   **Afzel.**
Agardh, Carl Adolf (Adolph) (1785-1859)   **C.Agardh**
Agardh, Jacob Georg (1813-1901)   **J.Agardh**
Agnew, Andrew David Quentin (1929-   )   **Agnew**
Ahti, Leena Hämet
    vide Hämet-Ahti, (Raija-)Leena   **Hämet-Ahti**
Aiton, William (1731-1793)   **Aiton**
Aizpuru Oiarbide, Iñaki (1956-   )   **Aizpuru**
Akasawa, Yoshyuki (1915-   )   **Akasawa**
Alain, Hermano (né Liogier Allut, Enrique Eugenio) (1916-   )   **Alain**
Albert, Abel (1836-1909)   **Albert**
Albov (Albow, Alboff), Nikolai Michailovich (1866-1897)   **Albov**
Alderwerelt van Rosenburgh, Cornelis Rugier (Rogier)
    Willem Karel van (1863-1936)   **Alderw.**
Alissova-Klobukova, Eugenija Nikolaevna (1889-1962)   **Aliss.**
Allan, Harry Howard (Barton) (1882- 1957)   **Allan**
Allchin, William Henry (c.1828-1891)   **Allchin**
Allen, Betty Eleanor Gosset (née Molesworth, B.E.G.) (1913-   )   **B.M.Allen**
Allioni, Carlo Ludovico (1728-1804)   **All.**
Almeida, Joseph Francis Raphael d'
    vide d'Almeida, J.F.R.   **d'Almeida**
Almeida, Marselin Rusario (1939-   )   **M.R.Almeida**
Almeida, Maria Teresa de (1940-   )   **M.T.Almeida**
Almeida, Sarah Marselin (née Saramma Verghese) (1940-   )   **S.M.Almeida**
Alston, Arthur Hugh Garfit (1902-1958)   **Alston**
Alt, Karen Susan (later Grant, Karen Alt) (1935-   )   **K.S.Alt**
Alverson, Edward R. (fl.1989)   **E.R.Alverson**
Amann, Johann (né Kurz, Wilhelm Sulpiz) (1834-1878)   **Amann**
Amano, Tetsuo (1912-   )   **T.Amano**
Amaral Franco, João Manuel António Paes do
    vide Franco, J.M.A. Paes do Amaral   **Franco**

Ammal, L. Sankari (fl. 1988) ✳	**Ammal**
Amman, Johann [latine Ammanus] (1707-1741)	**Amman**
Amoroso, Victor Bucad (1953-)	**Amoroso**
Amstutz, (Valentine) Erika (née Zwirbulis, E.) (1926-)	**Amstutz**
Anderson, Jacob Peter (1874-1953)	**J.P.Anderson**
Anderson, William Russell (1942-)	**W.R.Anderson**
Andersson, (Bengt) Lennart (1948-)	**L.Andersson**
Andersson, Nils Johan (1821-1880)	**Andersson**
Andrade Lima, Arturo Dárdano de (1919-1981)	**Andrade Lima**
Andrásovszky (Andrászovszky), Jószef (Josef) (1889-1943)	**András.**
André, Édouard-François (1840-1911)	**André**
Andrews, Elisabeth Gale (later Hooper, Elisabeth	
Andrews (1960-)	**E.G.Andrews**
Andrews, Spencer Bruce (1931-)	**S.B.Andrews**
Angely, João Alberto (1917-)	**Angely**
Ångström, Johan (1813-1879)	**Ångstr.**
Anthony, Nicola Christine (née Parkman, N.C.) (1957-)	**N.C.Anthony**
Appleby, Samuel (1806-1870)	**Appleby**
Arbeláez, Alba Luz (1965-)	**Arbeláez**
Arbeláez, Enrique Pérez	
vide Pérez Arbeláez, E.	**Pérez Arbel.**
Arcangeli, Giovanni (1840-1921)	**Arcang.**
Armstrong, John Beattie (1850-1926)	**J.B.Armstr.**
Arnott, George Arnott (Arnold) Walker (1799-1868)	**Arn.**
Arthur, Joseph Charles (1850-1942)	**Arthur**
Arya, Bhanwar Singh (1938-)	**B.S.Arya**
Ascherson, Paul (Friedrich August) (1834-1913)	**Asch.**
Askerov, Ajdyn Musaogly (1948-)	**Askerov**
Asplund, Erik (1888-1974)	**Aspl.**
Aswal, Bachan Singh (1948-)	**Aswal**
Atehortúa, Lucía (1949-)	**Atehortúa**
Atehortúa, Lucía Garcés	
vide Atehortúa, L.	**Atehortúa**
Atkinson, William Sackston (1821-1875)	**Atk.**
Attinger, Ernst (1890-1978)	**Attinger**
Aubert du Petit-Thouars, Louis-Marie Aubert	
vide Du Petit-Thouars, L.-M.A. Aubert	**Thouars**
Aublet, Jean Baptiste Christophe, Fusée (1720-1778)	**Aubl.**
Avetta, Carlo (1861-1941)	**Avetta**
Azevedo de Menezes, Carlos	
vide Menezes, C. Azevedo de	**Menezes**
Babington, Charles Cardale (1808- 1895)	**Bab.**
Bachelot de la Pylaie, Auguste Jean Marie (1786-1856)	**Bach.Pyl.**
Backer, Cornelis Andries (1874-1963)	**Backer**
Backhouse, James (pater) (1794-1869)	**Backh.**

Backhouse, James (filius) (1825-1890)	**Backh.f.**
Badré, Frederic Jean (1937-)	**Badré**
Baenitz, Karl (Carl) Gabriel (1837-1913)	**Baen.**
Baesecke, (Karl Heinrich Johannes) Paul (1877-1929)	**Baesecke**
Bailey, Frederick Manson (1827-1915)	**F.M.Bailey**
Bailey, Liberty Hyde (1858-1954)	**L.H.Bailey**
Baillon, Henri Ernest (1827-1895)	**Baill.**
Bain, John (1815-1903)	**J.Bain**
Baker, Edmund Gilbert (1864-1949)	**Baker f.**
Baker, John Gilbert (1834-1920)	**Baker**
Balakrishnan, Nambiyath Puthansurayil (1935-)	**N.P.Balakr.**
Balbis, Gioanni (Giovanni) Battista (1765-1831)	**Balb.**
Balfour, Isaac Bayley (1853-1922)	**Balf.f.**
Ballard, Francis (1896-1976)	**F.Ballard**
Ballard, Harvey Eugene Jr. (1958-)	**H.E.Ballard**
Bange, Anthelme-Jean (1896-1950)	**Bange**
Bär, Johannes (1877-1957)	**Bär**
Barber, Horace Newton (1914-1971)	**H.N.Barber**
Barbey, William (1842-1914)	**Barbey**
Barbey-Boissier, William	
vide Barbey, W.	**Barbey**
Barbosa Rodrigues, João (1842-1909)	**Barb.Rodr.**
Barnes, James Martinsdale (1814-1890)	**J.M.Barnes**
Barnes, T.M. (fl. 1860s)	**T.M.Barnes**
Barnola, Joaquín María de (1870-1925)	**Barnola**
Baroni, Eugenio (1865-1943)	**Baroni**
Barrelier, Jacques [latine Barrelierus] (1606-1673)	**Barr.**
Barrington, David Stanley (1948-)	**Barrington**
Bartlett, Harley Harris (1886-1960)	**Bartlett**
Bäsecke, (Karl Heinrich Johannes) Paul	
vide Baesecke, (K.H.J.) P.	**Baesecke**
Bastard, Toussaint (1784-1846)	**Bastard**
Basu, Somen Kumar (1951-)	**S.K.Basu**
Batard, Toussaint	
vide Bastard, T.	**Bastard**
Batarda Fernandes, Rosette Mercedes Saraiva	
vide Fernandes, R.M. Saraiva Batarda	**R.Fern.**
Batarda, Rosette Mercedes Saraiva (later Fernandes, R.M.) (1916-)	**Batarda**
Batsch, August Johann Georg Carl (Karl) (1761-1802)	**Batsch**
Battandier, Jules Aimé (1848-1922)	**Batt.**
Bauer, Franz (Francis) Andreas (1758-1840)	**F.A.Bauer**
Bauhin, Caspar (Kaspar; Gaspard) [latine C.Bauhinius]	
(1560-1624)	**C.Bauhin**
Bauhin, Johann (Jean) [latine J.Bauhinius] (1541-1612)	**J.Bauhin**
Baumann-Bodenheim, Marcel Gustav (fl. 1989)	**Baum.-Bod.**
Baumgarten, Johann Christian Gottlob (1765-1843)	**Baumg.**

Bautista, Hortensia Pousada (1949-)	**Bautista**
Beadle, Noel Charles William (1914-)	**N.C.W.Beadle**
Beaman, John Homer (1929-)	**Beaman**
Beaman, Reed Schiele (1961-)	**R.S.Beaman**
Beatson, Alexander (1759-1833)	**Beatson**
Beauverd, Gustave (1867-1942)	**Beauverd**
Beauvois, Ambroise Marie François Joseph Palisot de	
vide Palisot de Beauvois, A.M.F.J.	**P.Beauv.**
Beccari, Odoardo (1843-1920)	**Becc.**
Becherer, Alfred (1897-1977)	**Bech.**
Beck, Günther, von Mannagetta und Lërchenau	
vide Beck von Mannagetta und Lërchenau, G.	**Beck**
Beck, Lewis (Louis) Caleb (1798-1853)	**L.C.Beck**
Beck von Mannagetta und Lërchenau, Günther (1856-1931)	**Beck**
Beddome, Richard Henry (1830-1911)	**Bedd.**
Beekman, Willem Lubbertus (1895-)	**Beekman**
Béguinot, Augusto (1875-1940)	**Bég.**
Beitel, Joseph Michael (1952-1991)	**Beitel**
Bélanger, Charles Paulus (1805-1881)	**Bél.**
Bell, Peter Robert (1920-)	**P.R.Bell**
Bellair, Georges Adolphe (1860-1939)	**Bellair**
Bellardi, Carlo Antonio Lodovico (1741-1826)	**Bellardi**
Bellynck, Auguste Alexis Adolphe Alexandre (1814-1877)	**Bellynck**
Benedict, James Everard Jr. (1885-1969)	**J.E.Benedict**
Benedict, Ralph Curtiss (1883-1965)	**Benedict**
Benham, Dale Maurice (1957-)	**D.M.Benham**
Benl, Gerhard (1910-)	**Benl**
Bennert, Herbert Wilfried (1945-)	**Bennert**
Bennet, Sigamony Stephans Richardson (1940-)	**Bennet**
Bennett, John (Joannes) Joseph (1801-1876)	**Benn.**
Bentham, George (1800-1884)	**Benth.**
Beppu, Minoru (c.1944-)	**M.Beppu**
Berchtold, Bedřicha (Friedrich) Wssemjra von (1781-1876)	**Bercht.**
Berg, Maria Elizabeth van den (1946-)	**M.E.Berg**
Bergdolt, Ernst (1902-1948)	**Bergdolt**
Berger, Chr. J. (1724-1789)	**C.J.Berger**
Bergius, Bengt (Benedictus) (1723-1784)	**Bergius**
Bergius, Peter Jonas (1730-1790)	**P.J.Bergius**
Bernhardi, Johann Jakob (1774-1850)	**Bernh.**
Bernoulli, Karl (Carl) Gustav (1834-1878)	**Bernoulli**
Bertero, Carlo Luigi Giuseppe (1789-1831)	**Bertero**
Berthelot, Sabin (1794-1880)	**Berthel.**
Berthet, Paul (1933-)	**Berthet**
Bertoloni, Antonio (1775-1869)	**Bertol.**
Bertsch, Karl (1878-1965)	**Bertsch**
Betche, Ernst (1851-1913)	**Betche**

4

Bevis, John ("Ino") (1781-1876)	**Bevis**
Beyrich, Heinrich Karl (1796-1834)	**Beyr.**
Bhardwaja, Triloki Nath (1933-)	**Bhardwaja**
Bhavanandan, Kochu Veloo (1938-) *	**Bhavanan.**
Bhu, Indira (née Goswami, I.) (1962-)	**Bhu**
Bicknell, Clarence (1842-1918)	**C.Bicknell**
Bidin, Abdul Aziz (1948-)	**Bidin**
Bieberstein, Friedrich August Marschall von	
vide Marschall von Bieberstein, F.A.	**M.Bieb.**
Bierhorst, David William (1924-1966)	**Bierh.**
Billet, Albert Paul (1858-1915)	**Billet**
Binnendijk, Simon (1821-1883)	**Binn.**
Bir, Sarmukh Singh (1929-)	**Bir**
Bischoff, Gottlieb Wilhelm (Theophilus Guilielmus) (1797-1854)	**Bisch.**
Bishop, Luther Earl (1943-1991)	**L.E.Bishop**
Bissell, Charles Humphrey (1857-1925)	**Bissell**
Biswas, Anjali (née Das, A.) (1947-)	**A.Biswas**
Biswas, Kalipada P. (1899-1969)	**Biswas**
Biswas, Mantu Charan (1943-)	**M.C.Biswas**
Bitter, (Friedrich August) Georg (1873-1927)	**Bitter**
Bizzarri, Maria Paola (1937-)	**Bizzarri**
Black, John McConnell (1855-1951)	**J.M.Black**
Blake, Sidney Fay (1892-1959)	**S.F.Blake**
Blakiston, Thomas Wrigt (1832-1891)	**Blakiston**
Blanchard, William Henry (1850-1922)	**Blanch.**
Blanco, Francisco Manuel (1778-1845)	**Blanco**
Blanford, Henry Francis (1834-1893)	**Blanf.**
Blasdell, Robert Ferris (1929-)	**Blasdell**
Blatter, Ethelbert (1877-1934)	**Blatt.**
Blomquist, Hugo Leander (1885-1964)	**H.L.Blomq.**
Blume, Carl (Karl) Ludwig von (1796-1862)	**Blume**
Blytt, Axel Gudbrand (1843-1898)	**A.Blytt**
Bobrov, Andrej Eugenievich (1936-)	**A.E.Bobrov**
Boccone, Paolo (later Silvio) (1633-1703)	**Bocc.**
Böckel (Boeckel), Godwin (fl. 1853-1867)	**Böckel**
Boehmer, Georg Rudolf (Rudolph) (1723-1803)	**Boehm.**
Boer, Jan Gerard Wessels	
vide Wessels Boer, J.G.	**Wess.Boer**
Boerner, Carl (Karl) Julius Bernhard	
vide Börner, C.(K.) J.B.	**Börner**
Boggiani, Oliviero (1859-1933)	**Boggiani**
Böhmer, Georg Rudolf (Rudolph)	
vide Boehmer, G.R.	**Boehm.**
Boissier de la Croix de Sauvages, François	
vide Sauvages, F. Boissier de la Croix de	**Sauvages**
Boissier, Pierre Édmond (1810-1885)	**Boiss.**

Boivin, (Joseph Robert) Bernard (1916-1985)   **B.Boivin**
Bojer, Wenceslas (Wenceslaus, Wencelaus, Wenzel) (1797-1856)   **Bojer**
Boldingh, Isaäc (1879-1938)   **Bold.**
Bole, Pritamlal Vijayshankar (1920-   )   **Bole**
Bolle, Carl (Karl) August (1821-1909)   **Bolle**
Bolòs i Capdevila, Oriol de
   vide Bolòs, O. de   **O.Bolòs**
Bolòs, Oriol de (1924-   )   **O.Bolòs**
Bolton, James (1758-1799)   **Bolton**
Bolzon, Pio (1867-1940)   **Bolzon**
Bommer, Jean-Édouard (1829-1895)   **J.Bommer**
Bonaparte, Roland Napoléon (1858-1924)   **Bonap.**
Bonati, Gustave Henri (1873-1927)   **Bonati**
Bongard, August Gustav Heinrich von (1786-1839)   **Bong.**
Bonner, Charles Edmond Bradlaugh (1915-1976)   **Bonner**
Bonpland, Aimé Jacques Alexandre (né Goujaud, A.J.A.) (1773-1858)   **Bonpl.**
Boom, Bryan Morley (1954-   )   **B.M.Boom**
Borbás, Vinczé (Vincent, Vince, Vincenz) von (1844-1905)   **Borbás**
Borckhausen, Moritz (Moriz) Balthasar
   vide Borkhausen, M.B.   **Borkh.**
Bordonneau, Michel (fl. 1989)   **Bordonn.**
Boreau, Alexandre (1803-1875)   **Boreau**
Borg, John (1873-1945)   **Borg**
Borkhausen, Moritz (Moriz) Balthasar (1760-1806)   **Borkh.**
Borkowski, Roman (1882-   ?   )   **Bork.**
Börner, Carl (Karl) Julius Bernhard (1880-1953)   **Börner**
Borssum Waalkes, Jan van (1922-1985)   **Borss.Waalk.**
Bory de Saint-Vincent, Jean Baptiste Georges Geneviève
   Marcellin (1778-1846)   **Bory**
Bory, Jean Baptiste Georges Geneviève Marcellin de Saint-Vincent
   vide Bory de Saint-Vincent, J.B.G.G.M.   **Bory**
Borzì, Antonino (1852-1921)   **Borzì**
Bosc, Louis Auguste (Augustin) Guillaume (1759-1828)   **Bosc**
Bosch, Roelof Benjamin van den
   vide van den Bosch, R.B.   **Bosch**
Bosco, Roberto (1902-   )   **Bosco**
Bosman, Monica (Monique) Theresia Maria (1958-   )   **Bosman**
Bostock, Peter Dundas (1949-   )   **Bostock**
Boswell-Syme, John Thomas Irvine (né Syme, J.T.I.)
   vide Syme, J.T.I.   **Syme**
Bouché, Carl David (1809-1881)   **C.D.Bouché**
Boudrie, Michel (1950-   )   **Boudrie**
Boufford, David Edward (1941-   )   **Boufford**
Bowdich, Sarah (née Wallis, S.; remarried Lee, S.) (1791-1856)   **Bowdich**
Bowdich, Thomas Edward (1791-1824)   **T.E.Bowdich**
Bower, Frederick Orpen (1855-1948)   **Bower**

Bowie, James (1789-1869) — **Bowie**
Box, Harold Edmund (1898-?1972) — **Box**
Boydston, Kathryn (Kay) Estelle (1897-1988) — **Boydston**
Brackenridge, William Dunlop (1810-1893) — **Brack.**
Brade, Alexandre Curt (1881-1971) — **Brade**
Braggins, John Edward (1944-) — **Braggins**
Braithwaite, Anthony Forrester (1936-) — **A.F.Braithw.**
Braun, Alexander (Carl, Karl, Heinrich) (1805-1877) — **A.Braun**
Braun, Emma Lucy (1889-1971) — **E.L.Braun**
Brause, Guido Georg Wilhelm (1847-1922) — **Brause**
Bremekamp, Cornelis Eliza (Elisa) Bertus (1888-1984) — **Bremek.**
Brenan, John Patrick Micklethwait (1917-1985) — **Brenan**
Brewer, William Henry (1828-1910) — **W.H.Brewer**
Breyne (Breyn), Jacob (Jakob) [latine Breynius] (1637-1697) — **Breyne**
Brichan, James B. (fl. 1832-1845) — **Brichan**
Briquet, John Isaac (1870-1931) — **Briq.**
Brisseau de Mirbel, Charles François
    vide Mirbel, C.F. Brisseau de — **Mirb.**
Britten, James (1846-1924) — **Britten**
Britton, Donald Macphail (1923-) — **D.M.Britton**
Britton, Elizabeth Gertrude (née Knight, E.G.) (1858-1934) — **E.Britton**
Britton, Nathaniel Lord (1859-1934) — **Britton**
Broadhurst, Jean Alice (1873-1954) — **Broadh.**
Brockmann-Jerosch, Henryk (né Krzymowski, H., later
    Brockmann, H.) (1879-1939) — **Brockm.-Jer.**
Brockmüller, Hans Joachim Heinrich (1821-1882) — **Brockm.**
Bromhead, Edward French (1789-1855) — **Bromhead**
Brongniart, Adolphe Théodore (de) (1801-1876) — **Brongn.**
Brooks, Cecil Joslin (1875-1953) — **Brooks**
Brooks, Ralph Edward (1950-) — **R.E.Brooks**
Broun, Maurice (1906-) — **M.Broun**
Brown, Addison (1830-1913) — **A.Br.**
Brown, Clair Alan (1903-) — **C.A.Br.**
Brown, Donald Frederick McKenzie (1919-) — **D.F.Br.**
Brown, Elizabeth Dorothy (née Wuist, E.D.) (1880-1972) — **E.D.Br.**
Brown, Forest Buffen Harkness (1873-1954) — **F.Br.**
Brown, Robert (1773-1858) — **R.Br.**
Brown, Robert Neal Rudmose (1879-1957) — **R.N.R.Br.**
Browne, Patrick (1720-1790) — **P.Browne**
Brownlie, Garth (1923-1986) — **Brownlie**
Brownsey, Patrick John (1948-) — **Brownsey**
Bruce, James Garnett (III) (1944-) — **J.G.Bruce**
Brückner (Brueckner), Adolf Friedrich Albrecht (1781-1818) — **Brückn.**
Bruegger von Churwalden, Christian Georg
    vide Brügger, C.G. — **Brügger**
Brügger, Christian Georg (1833-1899) — **Brügger**

7

Bruhin, Pater Thomas von Aquin (Aquinas, né Gottfried) (1835-1896)   **Bruhin**
Brunton, Daniel Francis (1948-   )   **D.F.Brunt.**
Bryant, Truman Rai (1932-   )   **T.R.Bryant**
Bubani, Pietro (1806-1888)   **Bubani**
Buch, Christian Leopold von (1774-1853)   **Buch**
Buchanan, Francis (from 1818 Hamilton, F.) (1762-1829)   **Buch.-Ham.**
Buchheister, John C. (fl. 1903)   **Buchheist.**
Buchholz, John Theodore (1888-1951)   **J.Buchholz**
Buck, William Russell (1950-   )   **W.R.Buck**
Buckley, Samuel Botsford (1809-1884)   **Buckley**
Bueno, Rogerio Machado (1955-   )   **R.M.Bueno**
Bull, William (1828-1902)   **W.Bull**
Burchell, William John (1781-1863)   **Burch.**
Burck, William (1848-1910)   **Burck**
Burdet, Hervé Maurice (1939-   )   **Burdet**
Bureau, Louis Édouard (1830-1918)   **Bureau**
Burgess, Edward Sandford (1855-1928)   **E.S.Burgess**
Burkart, Arturo Erhardo (Erardo) (1906-1975)   **Burkart**
Burman (Burmann), Johannes (Jan) [latine Burmannus]
     (1706/07-1779)   **Burm.**
Burman, Nicolaas Laurens (Nicolaus Laurent) (1733-1793)   **Burm.f.**
Burmeister, Hermann Carl (Carlos) Conrad (Conrado) (1807-1892)   **Burmeist.**
Burnat, Émile (1828-1920)   **Burnat**
Burnett, Gilbert Thomas (1800-1835)   **Burnett**
Burnham, Stewart Henry (1870-1943)   **Burnham**
Burrows, John Eric (1950-   )   **J.E.Burrows**
Burrows, Sandra Margaret (née Schultz, S.M.) (1959-   )   **S.M.Burrows**
Burser, Joachim [latine Burserus] (1583-1639)   **Burser**
Busch, Nicolai Adolfowitsch (Nicolaj Adolfovich) (1869-1941)   **N.Busch**
Bush, Benjamin Franklin (1858-1937)   **Bush**
Butters, Frederic King (1878-1945)   **Butters**
Buysson, Robert du
     vide du Buysson, R.   **du Buysson**

Cabezudo Artero, Baltasar
     vide Cabezudo, B.   **Cabezudo**
Cabezudo, Baltasar (1946-   )   **Cabezudo**
Cadeval i Diars, Joan
     vide Cadevall y Diars, Juan   **Cadevall**
Cadevall y Diars, Juan (1846-1921)   **Cadevall**
Calder, James (Jim) Alexander (1915-1990)   **Calder**
Callé, Jean (1906-   )   **Callé**
Camerarius, Joachim II (1534-1598)   **Camer.**
Cameron, David (c.1787-1848)   **D.Cameron**
Campbell, Douglas Houghton (1859-1953)   **Campb.**
Camus, Josephine Margaret (later Rankin, J.M.) (1949-   )   **J.M.Camus**

Candolle, Augustin Pyramus de (1778-1841)	**DC.**
Capurro, Roberto Horacio (1910-)	**Capurro**
Carhart, Macy (fl. 1916)	**Carhart**
Cariot, Antoine (1820-1883)	**Cariot**
Carmichael, Dugald (1772-1827)	**Carmich.**
Carolin, Rogers Charles (1929-)	**Carolin**
Carrière, Élie-Abel (1818-1896)	**Carrière**
Carruthers, William (1830-1922)	**Carruth.**
Carse, Harry (1857-1930)	**Carse**
Carvalho e Vasconcellos, João de	
vide Vasconcellos, J. de Carvalho e	**Vasc.**
Caspary, (Johann Xaver, Xaber) Robert (1818-1887)	**Casp.**
Castillo, Amparo (1947-)	**A.Castillo**
Castillo Garcia, Amparo	
vide Castillo, A.	**A.Castillo**
Catalán, Pilar (1958-)	**Catalán**
Catalán Rodríguez, Pilar	
vide Catalán, P.	**Catalán**
Caumel, Jean Baptiste (later Héribaud-Joseph, Frère)	
vide Héribaud-Joseph, Frère	**Hérib.**
Cavanilles, Antonio Joseph (José) (1745-1804)	**Cav.**
Cavanilles Palop, Antonio Joseph (José)	
vide Cavanilles, A.J.	**Cav.**
Cei, Giuseppe (José) (1918-)	**Cei**
Celakovsky (Celakavský), Ladislav Frantisek (1864-1916)	**L.F.Celak.**
Celakovsky, Ladislav Josef (1834-1902)	**Celak.**
Cesalpino, Andrea [latine Caesalpinus] (1519-1603)	**Cesalpino**
Cesati, Vincenzo de (1806-1883)	**Ces.**
Chambers, Thomas Carrick (1930-)	**T.C.Chambers**
Chamisseau de Boncourt, Louis Charles Adelaïde	
vide Chamisso, L.K.A. von	**Cham.**
Chamisso, Ludolf Karl Adelbert von (1781-1838)	**Cham.**
Chandra, Prakash (1937-)	**P.Chandra**
Chandra, Subhash (1943-)	**S.Chandra**
Chang, Chao-Chien (1900-1972)	**C.C.Chang**
Chang, Jin-Lun (1960-)	**J.L.Chang**
Chapman, Alvan (Alvin) Wentworth (1809-1899)	**Chapm.**
Chassagne, Maurice (1880-1960)	**Chass.**
Chatenier, Constant (1849-1926)	**Chatenier**
Chaubard, Louis Athanase (Anastase) (1785-1854)	**Chaub.**
Chauvin, François Joseph (1797-1859)	**Chauv.**
Cheeseman, Thomas Frederic (1846-1923)	**Cheeseman**
Chen, Feng-Hwai (Feng-Huai) (1900-1993)	**F.H.Chen**
Chen, Jing-Fu (Jian-Fu)	
vide Cheng, Jing-Fu	**J.F.Cheng**
Chen, Katherine Lim (fl. 1964)	**Kath.Chen**

9

Chenevard, Paul (1839-1919)	**Chenevard**
Cheng, Jing-Fu (Jian-Fu) (1927-)	**J.F.Cheng**
Cheng, Wan-Chun (1904-1983) *	**W.C.Cheng**
Cheng, Xiao (1957-)	**X.Cheng**
Cherler, Johann Heinrich [latine Cherlerus] (1570-1610)	**Cherler**
Chevalier, Auguste Jean Baptiste (1873-1956)	**A.Chev.**
Chevallier, François Fulgis (1796-1840)	**Chevall.**
Chien, Chia-Chu	
vide Chien, Jia-Jü	**J.J.Chien**
Chien, Jia-Jü (Jia-Ju) (1922-1984)	**J.J.Chien**
Chien, Sung-Shu (Chung-Shu) (1883-1965)	**S.S.Chien**
Chin, Jen-Chang	
vide Ching, Ren-Chang	**Ching**
Ching, Ren-Chang (1898-1986)	**Ching**
Chinnock, Robert James (1943-)	**Chinnock**
Chiovenda, Emilio (1871-1941)	**Chiov.**
Chiu, Pao-Lin (Pao-Ling) (1936-)	**P.L.Chiu**
Chiu, Pei-Hsi	
vide Chiu, Pei-Shi	**P.S.Chiu**
Chiu, Pei-Shi (1919-1987)	**P.S.Chiu**
Chiu, Pei-Xi	
vide Chiu, Pei-Shi	**P.S.Chiu**
Choi, Hong-Kun (Hong-Keun) (1952-)	**H.K.Choi**
Chou, Ruth Chen-Ying (fl. 1947)	**R.C.Y.Chou**
Chowdhery, Harsh J. (1949-)	**H.J.Choudhery**
Chowdhury, Nira Pad (1911-)	**N.P.Chowdhury**
Chowdhury, S. (fl. 1969-1972)	**S.Chowdhury**
Christ, (Konrad) Hermann (Heinrich) (1833-1933)	**Christ**
Christensen, Carl Frederik Albert (1872-1942)	**C.Chr.**
Christiansen, Albert (? -1917)	**A.Christ.**
Christmann, Gottlieb Friedrich (1752-1836)	**Christm.**
Christophersen, Erling (1898-)	**Christoph.**
Chrysler, Mintin Asbury (1871-1963)	**Chrysler**
Chu, Ve-Ming	
vide Chu, Wei-Ming	**W.M.Chu**
Chu, Wei-Ming (1930-)	**W.M.Chu**
Chun, Woon-Young (1894-1971)	**Chun**
Chun, Yung-Ho	
vide Chung, Yung-Ho	**Y.H.Chung**
Chung, Yung-Ho (1924-)	**Y.H.Chung**
Cinq-Mars, Lionel (1919-1973)	**Cinq-Mars**
Clairville, Joseph Philippe de (1742-1830)	**Clairv.**
Clapham, Abraham (fl. 1860s-1870s)	**Clapham**
Clarke, Charles Baron (1832-1906)	**C.B.Clarke**
Clarkson, Edward Hale (1866-1934)	**Clarkson**
Clausen, Robert Theodore (1911-1981)	**R.T.Clausen**

Clayton, John (1686-1773)	**J.Clayton**
Clemente y Rubio, Simon de Rojas (Roxas) (1777-1827)	**Clemente**
Clemesha, Stephen Chapman (1942-)	**Clemesha**
Clifford, Harold Trevor (1927-) *	**Clifford**
Clowes, Frederic (fl. 1850s-1860s)	**Clowes**
Clusius, Carolus	
vide L'Escluse, C. de	**Clus.**
Clute, Willard Nelson (1869-1950)	**Clute**
Co, Leonardo Legaspi (1953-)	**Co**
Cockayne, Leonard (Leonhard, C.) (1855-1934)	**Cockayne**
Cockerell, Theodore Dru Alison (1866-1948)	**Cockerell**
Cody, William James (1922-)	**Cody**
Coffin, R.L. (fl. 1940)	**Coffin**
Cohn, Ferdinand (Julius) (1828-1898)	**Cohn**
Colenso, (John) William (1811-1899)	**Colenso**
Colla, Luigi (Aloysius) (1766-1848)	**Colla**
Colmeiro, Miguel (1816-1901)	**Colmeiro**
Colmeiro y Penido, Miguel	
vide Colmeiro, M.	**Colmeiro**
Colomb, Marie Louis Georges (1856-)	**Colomb**
Commerson, Philibert (1727-1773)	**Comm.**
Comolli, Giuseppe (1780-1849)	**Comolli**
Compton, Robert Harold (1886-1979)	**Compton**
Conant, David Stoughton (1949-)	**D.S.Conant**
Constantine, Jean (née Vesey, J.) (1930-) *	**Constantine**
Conzatti, Casiano (Cassiano) (1862-1951)	**Conz.**
Cook, Orator Fuller (1867-1949)	**O.F.Cook**
Cooper-Driver, Gillian (1936-)	**Cooper-Driver**
Copeland, Edwin Bingham (1873-1964)	**Copel.**
Corda, August Karl Joseph (1809-1849)	**Corda**
Cordemoy, Eugène, Jacob de (1835-1911)	**Cordem.**
Corley, Hugh Vanner (1914-)	**Corley**
Cornut, Jacques Philippe [latine Cornutus] (1606-1651)	**Cornut**
Corradi, (Bartolomeo Giacomo) Rinaldo (1897-1976)	**Corradi**
Correll, Donovan Stewart (1908-1983)	**Correll**
Cosentini, Ferdinando (1769-1840)	**Cosent.**
Cosson, Ernest Saint-Charles (1819-1889)	**Coss.**
Coulter, John Merle (1851-1928)	**J.M.Coult.**
Coutinho, António Xavier, Pereira (1851-1939)	**Cout.**
Coville, Frederick Vernon (1867-1937)	**Coville**
Crabbe, James Albert (1914-)	**Crabbe**
Crane, Fern Ward (1906-)	**Crane**
Cranfill, Raymond Benton (1958-)	**Cranfill**
Crawford, Lloyd C. (fl. 1951)	**Crawford**
Cremers, Georges (1936-)	**Cremers**
Cretzoiu, Paul (1909-1946)	**Cretz.**

Croft, James R. (1951-)	**J.R.Croft**
Cronk, Quentin C.B. (fl. 1980)	**Cronk**
Cronquist, Arthur John (1919-1992)	**Cronquist**
Crookes, Marguerite Winifred (1895-1991)	**Crookes**
Crosby, Marshall Robert (1943-)	**Crosby**
Crossfield, J. (fl. 1860s)	**Crossf.**
Croxall, John Patrick (1946-)	**Croxall**
Cubas Domínguez, Paloma	
vide Cubas, P.	**Cubas**
Cubas, Paloma (1954-)	**Cubas**
Cufodontis, Georg (Giorgio) (1896-1974)	**Cufod.**
Cunningham, Allan (1791-1839)	**A.Cunn.**
Currey, Frederick (1819-1881)	**Curr.**
Cusick, Allison Willmont (1941-)	**Cusick**
Czerepanov, Sergei Kirillovich (1921-1995)	**Czerep.**
d'Almeida, Joseph Francis Raphael (1891-1949)	**d'Almeida**
d'Orbigny, (Alcide) Charles Victor Dessalines	
vide Orbigny, (A.) C.V.D. d'	**Orb.**
D'Souza, Maria I.C. (fl. 1995)	**D'Souza**
d'Urville, Jules Sébastien César Dumont	
vide Dumont d'Urville, J.S.C.	**d'Urville**
da Silva Ferreira Sampaio, Gonçalo António	
vide Sampaio, G.A. da Silva Ferreira	**Samp.**
Dale, Elisabeth (fl. 1901)	**E.Dale**
Dalibard, Thomas François (1703-1779)	**Dalib.**
Dalla Torre, Karl (Carl) Wilhelm von (1850-1928)	**Dalla Torre**
Dalla Torre von Thurnberg-Sternhoff, Karl (Carl) Wilhelm von	
vide Dalla Torre, K.(C.)W. von	**Dalla Torre**
Dalzell, Nicol (Nicolas) Alexander (1817-1878)	**Dalzell**
Damazio, Léonidas Botelho (1854-1922)	**Damazio**
Damboldt, Jürgen (1937-1978)	**Damboldt**
Dangeard, Pierre Clement Augustin (1862-1947)	**P.A.Dang.**
Däniker, Albert Ulrich (1894-1957)	**Däniker**
Darnaedi, Dedy (1952-)	**Darnaedi**
Das, Anjali (later Biswas, A.) (1947-)	**A.Das**
Das, Silpi (fl.1995)	**S.Das**
Davenport, George Edward (1833-1907)	**Davenp.**
de Almeida, Maria Teresa	
vide Almeida, M.T. de	**M.T.Almeida**
de Andrade Lima, Arturo Dárdano	
vide Andrade Lima, A.D. de	**Andrade Lima**
de Barnola, Joaquín María	
vide Barnola, J.M. de	**Barnola**
de Beauvois, Ambroise Marie François Joseph Palisot	
vide Palisot de Beauvois, A.M.F.J.	**P.Beauv.**

de Bolòs, Oriol	
vide Bolòs, O. de	**O.Bolòs**
de Candolle, Augustin Pyramus	
vide Candolle, A.P. de	**DC.**
de Carvalho e Vasconcellos, João	
vide Vasconcellos, J. de Carvalho e	**Vasc.**
de Cesati, Vincenzo	
vide Cesati, V. de	**Ces.**
de Clairville, Joseph Philippe	
vide Clairville, J.Ph. de	**Clairv.**
de Cordemoy, Eugène Jacob	
vide Cordemoy, E. Jacob de	**Cordem.**
de Ezcurdia, Luis	
vide Ezcurdia, L. de	**Ezcurdia**
de Freycinet, (Henri) Louis Claude de Saulces	
vide Freycinet, (H.) L.C.d.S.	**Freyc.**
de Garsault, François Alexandre Pierre	
vide Garsault, F.A.P. de	**Garsault**
de Joncheere, Gerardus Johannes Pieter (1909-1989)	**de Jonch.**
de Jussieu, Antoine Laurent	
vide Jussieu, A.L. de	**Juss.**
de Jussieu, Bernard	
vide Jussieu, B. de	**B.Juss.**
de l'Escluse (de l'Ecluse), Charles	
vide L'Escluse, C. de	**Clus.**
de la Peirouse, Philippe Picot	
vide Lapeyrouse, P. Picot de	**Lapeyr.**
de la Pylaie, August Jean Marie Bachelot	
vide Bachelot de la Pylaie, A.J.M.	**Bach.Pyl.**
de la Sota, Elias Ramón (1932-)	**de la Sota**
de Labillardière, Jacques Julien Houtton	
vide Labillardière, J.J.H. de	**Labill.**
de Lagasca y Segura, Mariano	
vide Lagasca y Segura, M. de	**Lag.**
de Lamarck, Jean Baptiste Antoine Pierre de Monnet	
vide Lamarck, J.B.A.P. de Monnet de	**Lam.**
De Langhe, Joseph Edgard (1907-)	**De Langhe**
de Lapeyrouse, Philippe Picot	
vide Lapeyrouse, P. Picot de	**Lapeyr.**
de Litardière, René Verriet	
vide Litardière, R.V. de	**Litard.**
de Lobel, Mathias	
vide Lobel, M. de	**Lobel**
de Loureiro, João	
vide Loureiro, J. de	**Lour.**

13

de Marchesetti, Carlo (Carl von)
    vide Marchesetti, C. de   **Marches.**
de Menezes, Carlos Azevedo
    vide Menezes, C.A.   **Menezes**
de Mirbel, Charles François Brisseau
    vide Mirbel, C.F. Brisseau de   **Mirb.**
de Monnet de Lamarck, Jean Baptiste Antoine Pierre
    vide Lamarck, J.B.A.P. de Monnet de   **Lam.**
de Necker, Noel Martin Joseph
    vide Necker, N.M.J. de   **Neck.**
De Notaris, Giuseppe (Josephus) (1805-1877)   **De Not.**
de Ortega, Casimiro Gómez
    vide Gómez (de) Ortega, C.   **Ortega**
de Resende-Pinto, Manuel Cabral
    vide Resende-Pinto, M.C. de   **Res.-Pinto**
de Rey-Pailhade, Constantin
    vide Rey-Pailhade, C. de   **Rey-Pailh.**
de Rezende-Pinto, Manuel Cabral
    vide Resende-Pinto, M.C. de   **Res.-Pinto**
de Saint-Hilaire, Auguste (Augustin) François César Prouvençal
    vide Saint-Hilaire, A.F.C. Prouvençal de   **A.St.-Hil.**
de Sampaio, Alberto José
    vide Sampaio, A.J. de   **A.Samp.**
de Sauvages, François Boissier de la Croix
    vide Sauvages, F. Boissier de la Croix de   **Sauvages**
de Savigny, (Marie) Jules César Lélorgne
    vide Savigny, (M.) J.C. Lélorgne de   **Savigny**
de Schoenefeld, Wladimir
    vide Schoenefeld, W. de   **Schoenef.**
De Toni, Giovanni Battista (1864-1924)   **De Toni**
de Visiani, Roberto (1800-1878)   **Vis.**
De Vol, Charles Edward (1903-1989)   **De Vol**
de Vriese, Willem Hendrik (1806-1862)   **de Vriese**
de Waha Baillonville, T.
    vide Waha Baillonville, T. de   **Waha**
De Wildeman, Émile August(e) Joseph (1866-1947)   **De Wild.**
Deakin, Richard (1808-1873)   **Deakin**
Deb, Debendra Bijoy (1924-   )   **Deb**
Decaisne, Joseph (1807-1882)   **Decne.**
Decandolle, Augustin Pyramus
    vide Candolle, A.P. de   **DC.**
Decken, Carl Claus von der (1833-1865)   **Decken**
Deferrari, Amelia Marta (1945-   )   **Deferrari**
Degen, Árpád von (1866-1934)   **Degen**
Degener, Irmgard (Isa) (née Hansen, I.) (1924-   )   **I.Deg.**
Degener, Otto (1899-1988)   **O.Deg.**

Delile, Alire Raffeneau (1778-1850) **Delile**
Demaret, Fernand Mathieu Hubert (1911-   ) **Demaret**
Demiriz, Hüsnü (1920-   ) **Demiriz**
Derrick, Lewis Norman (1948-   ) **L.N.Derrick**
Desfontaines, René Louiche (1750-1833) **Desf.**
Desmazières, Jean Baptiste Henri Joseph (1786-1862) **Desm.**
Desrousseaux, Louis Auguste Joseph (1753-1838) **Desr.**
Desvaux, Nicaise Auguste (Augustin Nicaise) (1784-1856) **Desv.**
Devi, Kamla (fl. 1969) **K.Devi**
Dhir, K.K. (1937-   ) ✳ **Dhir**
Díaz Gonzáles, Tomás Emilio
    vide Díaz, T.E. **T.E.Díaz**
Díaz, Tomás Emilio (1949-   ) **T.E.Díaz**
Dickason, Frederick Garrett (1904-   ) **Dickason**
Dickson, James (Jacobus) (1738-1822) **Dicks.**
Diddell, Mary Blain (née Wallace, M.B.) (? -1962) **Diddell**
Diels, Friedrich Ludwig Emil (1874-1945) **Diels**
Diem, José (1899-1986) **Diem**
Diklić, Nikola (1925-   ) **Diklić**
Dillen, Johann Jacob (Jakob), [latine Dillenius]
    (1684-1747) **Dill.**
Dillwyn, Lewis Weston (1778-1855) **Dillwyn**
Ding, Zuo-Chao (1940-   ) **Z.C.Ding**
Dix, William Leroy (1875- ?) **Dix**
Dixit, Ram Das (1942-   ) **R.D.Dixit**
do Amaral Franco, João Manuel António Paes
    vide Franco, J.M.A. Paes do Amaral **Franco**
Dobbie, Herbert Boucher (1852-1940) **Dobbie**
Dodge, Raynal (1844-1918) **R.Dodge**
Dodoens, Rembert [latine Dodonaeus] (1517/18-1585) **Dodoens**
Doell, Johann Christoph (Johannes Christian)
    vide Döll, J.C. **Döll**
Doerfler, Ignaz
    vide Dörfler, I. **Dörfl.**
Döll, Johann Christoph (Johannes Christian) (1808-1885) **Döll**
Domin, Karel (1882-1953) **Domin**
Don, David (1799-1841) **D.Don**
Don, George (1798-1856) **G.Don**
Donk, Marinus Anton (1908-1972) **Donk**
Donnell, Smith F.
    vide Smith, F.D. **F.Donn.Sm.**
Donnell, Smith, John
    vide Smith, J.D. **Donn.Sm.**
Donselaar, Johannes van (1928-   ) **Donsel.**
Döpp, Walter (1901-1963) **Döpp**
Dörfler, Ignaz (1866-1950) **Dörfl.**

15

Dorn, Robert Donald (1942-)	**Dorn**
Dosch, Ludwig (fl. 1873-1888)	**Dosch**
Dostál, Josef (1903-)	**Dostál**
Dowell, Philip (1864-1936)	**Dowell**
Drake del Castillo, Emmanuel (1855-1904)	**Drake**
Druce, George Claridge (1850-1932)	**Druce**
Druery, Charles Thomas (1843-1917)	**Druery**
Dryander, Jonas Carlsson (1748-1810)	**Dryand.**
du Buysson, Robert (fl. 1888)	**du Buysson**
Du Petit-Thouars, Louis-Marie Aubert, Aubert (1758-1831)	**Thouars**
Duby, Jean Étienne (1798-1885)	**Duby**
Dudley, Theodore Robert (1936-)	**T.R.Dudley**
Duek, Jacobo Jack (1936-)	**Duek**
Dulac, Joseph (1827-1897)	**Dulac**
Dumont d'Urville, Jules Sébastian César (1790-1842)	**d'Urv.**
Dumortier, Barthélemy Charles Joseph (1797-1878)	**Dumort.**
Dunal, Michel Félix (1789-1856)	**Dunal**
Dunn, Stephen Troyte (1868-1938)	**Dunn**
Duperrey, Louis Isidore (1786-1865)	**Duperrey**
Durand, Théophile Alexis (1855-1912)	**T.Durand**
Durande, Jean François (1732-1794)	**Durande**
Durieu de Maisonneuve, Michel Charles (1796-1878)	**Durieu**
Dusén, Per Karl Hjalmar (1855-1926)	**Dusén**
Duthie, Augusta Vera (1881-1963)	**A.V.Duthie**
Dutilly, Arthème Antoine (1896-1973)	**Dutilly**
Dutra, João (1862-1939)	**Dutra**
Duval-Jouve, Joseph (1810-1883)	**Duval-Jouve**
Dyce, James Wood (1905-)	**Dyce**
Dyer, William Turner Thiselton- *vide* Thiselton-Dyer, W.T.	**Dyer**
Dzwonko, Zbigniew (1947-)	**Dzwonko**
Eames, Edwin Hubert (1865-1948)	**Eames**
Eastman, Helen (1863-)	**Eastman**
Eaton, Alvah Augustus (1865-1908)	**A.A.Eaton**
Eaton, Amos (1776-1842)	**Eaton**
Eaton, Daniel Cady (1834-1895)	**D.C.Eaton**
Eaton, Richard Jefferson (1890-1976)	**R.J.Eaton**
Edgeworth, Michael Pakenham (1812-1881)	**Edgew.**
Edwards, David Sydney (1948-)	**D.S.Edwards**
Edwards, Peter John (1947-)	**P.J.Edwards**
Edwards, Trevor John (1960-)	**T.J.Edwards**
Eggert, Heinrich Karl Daniel (1841-1904)	**Eggert**
Eggleston, Willard Webster (1863-1935)	**Eggl.**
Ehrhart, Jakob Friedrich (1742-1795)	**Ehrh.**
Ehrler, (Philipp) Anton (1872-1965)	**Ehrler**

Eichler, August Wilheim (1839-1887)	**Eichler**
Eiger, J. (fl. 1956)	**Eiger**
Ekman, Erik Leonard (1883-1931)	**Ekman**
Elmer, Adolph Daniel Edward (1870-1942)	**Elmer**
Elworthy, Charles, (1805- ?)	**Elworthy**
Emerson, Jean Kathryn (1950-)	**J.K.Emers.**
Emmott, Janet Irene (1938-)	**Emmott**
Emory, William Hemsley (1811-1887)	**Emory**
Ende, Willen Pieter van den (fl. 1823)	**Ende**
Endlicher, Stephan (Friedrich) Ladislaus (1804-1849)	**Endl.**
Engelmann, Georg (George) Theodor (1809-1884)	**Engelm.**
Engler, (Heinrich Gustav) Adolf (1844-1930)	**Engl.**
Enys, J.D. (1837-1912)	**Enys**
Eriksson, Ove Erik (1935-)	**O.E.Erikss.**
Eschweiler, Franz Gerhard (Franciscus Gerardus) (1796-1831)	**Eschw.**
Eseltine, Glen Parker Van *vide* Van Eseltine, G.P.	**Van Eselt.**
Espinosa Bustos, Marcial Ramón (1874-1959)	**Espinosa**
Ettingshausen, Constantin (Konstantin) von (1826-1897)	**Ettingsh.**
Evans, A. Murray (1932-)	**A.M.Evans**
Evans, Obed David (1889-1975)	**O.D.Evans**
Ewan, Joseph Andorfer (1909-)	**Ewan**
Exell, Arthur Wallis (1901-1993)	**Exell**
Ezcurdia, Luis de (fl. 1899)	**Ezcurdia**
Faden, Robert Bruce (1942-)	**Faden**
Fairbrothers, D.E. (fl. 1992)	**Fairbr.**
Falk, Heinz (1923-)	**H.Falk**
Farrant, Penelope Anne (1953-)	**P.A.Farrant**
Farrar, Donald Ray (1941-)	**Farrar**
Farwell, Oliver Atkins (1867-1944)	**Farw.**
Faull, Joseph Horace (1876-1961)	**Faull**
Faxon, Charles Edward (1846-1918)	**Faxon**
Fay, Alice Dickinson (née Awtrey, A.D.) (1926-)	**A.Fay**
Featherman, Americus (1822-c.1880)	**Featherm.**
Fedtschenko (Fedchenko), Boris Alexjewitsch (Alexeevich) (1872-1947)	**B.Fedtsch.**
Fedtschenko, Olga Alexandrowna (née Armfeld, O.A.) (1845-1921)	**O. Fedtsch.**
Fée, Antoine Laurent Apollinaire (1789-1874)	**Fée**
Feilberg, Jon (1944-)	**Feilberg**
Fenaroli, Luigi (1899-1980)	**Fen.**
Fendler, August (1813-1883)	**Fendler**
Ferguson, William (1820-1887)	**Ferguson**
Fernald, Merritt Lyndon (1873-1950)	**Fernald**

Fernandes

Fernandes, Rosette Mercedes
    (née Saraiva Batarda, R.M.) (1916-)   **R.Fern.**
Fernández Areces, María Pilar (1959-)   **Fern.Areces**
Fernández Prieto, José Antonio (1950-)   **Fern.Prieto**
Fiek, Emil (1840-1897)   **Fiek**
Field, Henry Claylands (1825-1912)   **Field**
Fielding, Henry Barron (1805-1851)   **Fielding**
Filarszky, Nándor (1858-1941)   **Fil.**
Fiori, Adriano (1865-1950)   **Fiori**
Fischer, Friedrich Ernst Ludwig von (Fedor Bogdanovic) (1782-1854) **Fisch.**
Fischer, Hugo (1865-1939)   **H.Fisch.**
Fliche, Paul Henri Maria Thérèse André (1836-1908)   **Fliche**
Florence, Jacques (1951-)   **Florence**
Flowers, Seville (1900-1968)   **Flowers**
Floyd, Frederick ('Fred') Gillan (1869-1941)   **Floyd**
Flynn, Timothy W. (1958-)   **Flynn**
Fomin, Aleksandr Vasilievich (1869-1935)   **Fomin**
Font Quer, Pio (Pius) (José Mariano) (1888-1964)   **Font Quer**
Font y(i) Quer, Pio (Pius) (José Mariano)
    vide Font Quer, P.(J.M.)   **Font Quer**
Forbes, Henry Ogg (1851-1932)   **H.O.Forbes**
Formánek, Eduard (1845-1900)   **Formánek**
Forsskål, Pehr (Peter) (1732-1763)   **Forssk.**
Forster, Johann Georg Adam (1754-1794)   **G.Forst.**
Fosberg, Francis Raymond (1908-1993)   **Fosberg**
Foster, H. Lincoln (1906-1989)   **H.L.Foster**
Fournier, Eugène Pierre Nicolas (1834-1884)   **E.Fourn.**
Fournier, Paul-Victor (1877-1964)   **P.Fourn.**
Fowler, Keith (1936-)   **K.Fowler**
Fowler, Robert Lawrence (1910-)   **Fowler**
Francé, Raoul Henrich (1874-1943)   **Francé**
Franchet, Adrien René (1834-1900)   **Franch.**
Francis, George William (1800-1865)   **Francis**
Franco, João Manuel António Paes do Amaral (1921-)   **Franco**
Frank, Albert Bernhard (1839-1900)   **A.B.Frank**
Franken, Nicolaas Antonius Petrus (1954-)   **Franken**
Franklin, John (1786-1847)   **Franklin**
Franzé, Reszö (Resso)
    vide Francé, Raoul Heinrich   **Francé**
Fraser, Patrick Neill (1830-1905)   **P.N.Fraser**
Fraser-Jenkins, Christopher Roy (1948-)   **Fraser-Jenk.**
Freiberg, Wilhelm (fl. 1911)   **Freiberg**
French, Charles (1840-1890)   **French**
Freycinet, (Henri) Louis Claude de Saulces de (1779-1842)   **Freyc.**
Fries, Elias Magnus (1794-1878)   **Fr.**
Fries, (Klas) Robert Elias (1876-1966)   **R.E.Fr.**

Fries, Theodor (Thore) Magnus (1832-1913)   **Th.Fr.**
Friesner, Ray Clarence (1894-1952)   **Friesner**
Fritsch, Karl (1864-1934)   **Fritsch**
Frivaldsky, Emerich (Imre)
     vide Frivaldszky von Frivald, E.(I.)   **Friv.**
Frivaldszky von Frivald, Emerich (Imre) (1799-1870)   **Friv.**
Frye, Theodore Christian (1869-1962)   **Frye**
Fu, Shu-Hsia (1916-1986)   **S.H.Fu**
Fuchs, Hans Peter (1928-　　)   **H.P.Fuchs**
Fuchs, Leonhart (Leonhard) [latine Fuchsius] (1501-1566)   **L.Fuchs**
Fuchs-Eckert, Hans Peter
     vide Fuchs, H.P.   **H.P.Fuchs**
Fusée Aublet, Jean Baptiste Christophe
     vide Aublet, J.B.C. Fusée   **Aubl.**
Futák, Ján (1914-1980)   **Futák**
Futó, Mihaly (1882-1929)   **Futó**

Galeotti, Henri Guillaume (1814-1858)   **Galeotti**
Gammie, George Alexander (1864-1935)   **Gammie**
Gandhi, Kancheepuram N. (1948-　　)   **Gandhi**
Gandoger, Michel (1850-1926)   **Gand.**
Garber, Abram Paschal (1838-1881)   **Garber**
García Caluff, Manuel (1945-　　)   **García Caluff**
García de López, Ivonne (fl. 1978)   **García de López**
García, Donato (1782-1855)   **D.García**
García Nogueruela, Donato
     vide García, D.   **D.García**
Gardner, George (1812-1849)   **Gardner**
Garsault, François Alexandre Pierre de (1691-1778)   **Garsault**
Gasparrini, Guglielmo (1804-1866)   **Gasp.**
Gastony, Gerald Joseph (1940-　　)   **Gastony**
Gätzi, Walter (1901-　　)   **Gätzi**
Gaudichaud, Charles
     vide Gaudichaud-Beaupré, C.   **Gaudich.**
Gaudichaud-Beaupré (-Beaupres), Charles (1789-1854)   **Gaudich.**
Gay, Claude (1800-1873)   **Gay**
Gay, Honor Jane (1964-　　) ✳   **H.J.Gay**
Gay, Jacques Étienne (1786-1864)   **J.Gay**
Geert, August(e) Van
     vide Van Geert, A.   **Van Geert**
Geevarghese, Kuzuvila Kurian (1943-　　)   **Geev.**
Geisenheyner, Franz Adolf Ludwig (1841-1926)   **Geisenh.**
Geissert, Fritz (Frédéric) (1923-　　)   **Geissert**
Gelert, Otto Christian (Kristian) Leonor (Laurits) (1862-1899)   **Gelert**
Gena, Chatur Bhuj (1947-　　)   **Gena**
Gennari, Patrizio (1820-1897)   **Gennari**

Georgi, Johann Gottlieb (1729-1802)	**Georgi**
Gepp, Anthony (Antony) (1862- 1955)	**A.Gepp**
Gérardin de Mirecourt, Sébastien	
vide Gérardin, S.	**Gérardin**
Gérardin, Sébastien (1751-1816)	**Gérardin**
Germain de Saint-Pierre, Jacques Nicolas Ernest (1815-1882)	**Germ.**
Gerrard, William Tyrer (? -1866)	**Gerrard**
Ghatak, Jagadananda (1928-)	**J.Ghatak**
Ghoryaninov, Paul Federowitsch	
vide Horaninow, P.F.	**Horan.**
Ghosh, Basabendra (1942-)	**B.Ghosh**
Ghosh, Ranjit Kumar (1941-)	**R.K.Ghosh**
Ghosh, Santi Ranjan (1942-)	**S.R.Ghosh**
Gibbs, Lilian Suzette (1870-1925)	**Gibbs**
Gibby, Mary (1949-)	**Gibby**
Gibelli, Giuseppe (1831-1898)	**Gibelli**
Giesenhagen, Karl (Carl) Friedrich Georg (1860-1928)	**Giesenh.**
Gilbert, Benjamin Davis (1835-1907)	**Gilbert**
Gilbert, Elizabeth Florence (1929-)	**E.F.Gilbert**
Gilibert, Jean-Emmanuel (1741-1814)	**Gilib.**
Gillespie, James Pottard (1931-)	**J.P.Gillespie**
Gilli, Alexander (1904-)	**Gilli**
Gilman, A.V. (fl. 1994)	**A.V.Gilman**
Giráldez Fernandez, Ximena	
vide Giráldez, X.	**Giráldez**
Giráldez, Ximena (1944-)	**Giráldez**
Girardet, Antonin (fl. 1891)	**Girardet**
Giudice, Gabriela Elena (1960-)	**Giudice**
Given, David Roger (1943-)	**Given**
Glassman, Sidney Frederick (1919-)	**Glassman**
Glaziou, Auguste François Marie (1828-1906)	**Glaz.**
Gleason, Henry Allan (1882-1975)	**Gleason**
Gleditsch, Johann Gottlieb (1714-1786)	**Gled.**
Glück, Christian Maximilian Hugo (1868-1940)	**Glück**
Gmelin, Johann Friedrich (1748-1804)	**J.F.Gmel.**
Gmelin, Samuel Gottlieb (1744-1774)	**S.G.Gmel.**
Goddijn, Wouter Adriaan (1884-1960)	**Goddijn**
Godet, Charles Henry (1797-1879)	**Godet**
Godron, Dominique Alexandre (1807-1880)	**Godr.**
Goebel, Karl Immanuel Eberhard von (1855-1932)	**K.I.Goebel**
Goeppert, Johann Heinrich Robert	
vide Göppert, J.H.R.	**Göpp.**
Goiran, Agostino (Augustin) (1835-1909)	**Goiran**
Goldie, John (1793-1886)	**Goldie**
Goldmann, Ignaz G. (1810-1848)	**Goldm.**

Golitsin (Golitzyn, Golicin), Sergey (Sergius)
    Vladimirovich (1897-1968) — **Golitsin**
Goltz, J.P. (fl. 1991) — **Goltz**
Gómez (de) Ortega, Casimiro (1740-1818) — **Ortega**
Gómez, Luis Diego (1944-   ) — **L.D.Gómez**
Gómez Pignataro, Luis Diego
    vide Gómez, L.D. — **L.D.Gómez**
Gómez-Laurito, Jorge (1947-   ) — **Gómez-Laur.**
Goodding, Leslie Newton (1880-1967) — **Goodd.**
Goode, John B. (fl. 1881) — **Goode**
Gopal, Brij (1944-   ) — **Gopal**
Göppert, Johann Heinrich Robert (1800-1884) — **Göpp.**
Gorianinov, Pavel Federowitsch
    vide Horaninov, P.F. — **Horan.**
Gorodkov (Gorodkow), Boris Nicolaevich (1890-1953) — **Gorodkov**
Goswami, Hit Kishore (1942-   ) — **Goswami**
Gouan, Antoine (1733-1821) — **Gouan**
Goy, Doris Alma (later Smith, D.A.) (1912-   ) — **Goy**
Gradaille, Josep Lluis (1946-   ) — **Gradaille**
Gradaille Tortella, Josep Lluis
    vide Gradaille, J.L. — **Gradaille**
Graebner, (Karl Otto Robert Peter) Paul (l871-1933) — **Graebn.**
Granié, Étienne Marcellin (later Frère Sennen)
    vide Sennen, Frère — **Sennen**
Granier-Blanc, Étienne Marcellin (later Frère Sennen)
    vide Sennen, Frère — **Sennen**
Grant, Karen Alt (née Alt, Karen Susan) (1935-   ) — **K.A.Grant**
Grant, Verne Edwin (1917-   ) — **V.E.Grant**
Graves, Edward Willis (1882-1936) — **E.W.Graves**
Graves, James Ansel (1828-1909) — **J.A.Graves**
Gray, Alan Maurice (1943-   ) — **A.M.Gray**
Gray, Asa (1810-1888) — **A.Gray**
Gray, Bruce (1939-   ) — **B.Gray**
Gray, Frederick William (1878- ?) — **F.W.Gray**
Gray, Samuel Frederick (1766-1828) — **Gray**
Grayum, Michael Howard (1949-   ) — **Grayum**
Green, Peter Shaw (1920-   ) — **P.S.Green**
Greene, Edward Lee (1843-1915) — **Greene**
Greene, Frank Cook (1886- ?) — **F.C.Greene**
Greenman, Jesse More (1867-1951) — **Greenm.**
Gremli, August(e) (1833-1899) — **Gremli**
Grenier-Blanc, Étienne Marcellin (later Frère Sennen)
    vide Sennen, Frère — **Sennen**
Grenier, Jean Charles Marie (1808-1875) — **Gren.**
Grether, David Frank (1920-   ) — **Grether**
Greuter, Werner Rodolfo (1938-   ) — **Greuter**

Greville, Robert Kaye (1794-1866)	**Grev.**
Griffith, William (1810-1845)	**Griff.**
Grimes, James Walter (1953-)	**J.W.Grimes**
Grinţescu, Ioan (1874-1963)	**I.Grinţ.**
Grisebach, August Heinrich Rudolf (Rudolph) (1814-1879)	**Griseb.**
Grisley, Gabriel (fl. 1661)	**Grisley**
Gronovius, Johan (Jan) Frederik (Frederic, Fredrik) (1686-1762)	**Gronov.**
Grossheim (Grossgeim), Alexander Alfonsovich (Alphonsovitch) (1888-1948)	**Grossh.**
Grout, Abel Joel (1867-1947)	**Grout**
Gruber, Calvin Luther (1864-1943)	**Gruber**
Grubov, Valery Ivanovich (1917-)	**Grubov**
Guadagno, Michele (1878-1930)	**Guadagno**
Gudoschnikov, Sergei Vasilevich (1916-)	**Gudoschn.**
Guédès, Michel (1942-1985)	**Guédès**
Gueldenstaedt, Johann Anton von (1745-1781)	**Gueldenst.**
Guérin, Joseph Xavier Benezet (1775-1850)	**Guérin**
Guétrot, M. (1873-1941)	**Guétrot**
Guettard, Jean Étienne (1715-1786)	**Guett.**
Guillaumin, André (1885-1974)	**Guillaumin**
Guillemin, Jean Baptiste Antoine (1796-1842)	**Guill.**
Guinea, Emilio (1907-1985)	**Guinea**
Guinea López, Emilio *vide* Guinea, E.	**Guinea**
Guo, Xiao-Si (1962-)	**X.S.Guo**
Gupta, Kedar Mal (1908-1987)	**K.M.Gupta**
Gureyeva (Gureeva), Irina (Irene) Ivanovna (1955-)	**Gureyeva**
Gurung, Vidya Laxmi (1940-)	**Gurung**
Gussone, Giovanni (1787-1866)	**Guss.**
Gutiérrez, Hermes Garces (1933-)	**H.G.Gut.**
Haberer, Joseph Valentine (1855-1925)	**Haberer**
Hagemann, Wolfgang (1929-)	**W.Hagemann**
Hagenah, Dale James (1908-1971)	**Hagenah**
Hagenah, Ethelda (1911-)	**E.Hagenah**
Hahne, August Hermann (1873-1942)	**Hahne**
Haines, Henry Haselfoot (1867-1945)	**Haines**
Halácsy, Eugen von (Eugène de) (1842-1913)	**Halácsy**
Hall, Carlotta (née Case, C.) (1880-1949)	**C.C.Hall**
Hall, Herman (Hermanus) Christiaan van (1801-1874)	**H.C.Hall**
Hall, James (1811-1898)	**J.Hall**
Hall, John Bartholomew (1932-1984)	**J.B.Hall**
Hall, John Walton (1918-)	**J.W.Hall**
Hallé, Nicolas (1927-)	**N.Hallé**
Haller, (Victor) Albrecht von [latine Hallerus] (1708-1777)	**Haller**

Halloy, Stephan (1953-)	**Halloy**
Hamann, Ole Jorgen (1944-)	**O.J.Hamann**
Hamet-Ahti, (Raija-)Leena (1931-)	**Hämet-Ahti**
Hamilton, Francis (né Buchanan, F.)	
vide Buchanan, F.	**Buch.-Ham.**
Hance, Henry Fletcher (1827-1886)	**Hance**
Handel-Mazzetti, Heinrich von (1882-1940)	**Hand.-Mazz.**
Handro, Osvaldo (1908-1986)	**Handro**
Hansen, Alfred (1925-)	**A.Hansen**
Hansen, Irmgard (later Degener, I.) (1924-)	**I.Hansen**
Hanstein, Johannes Ludwig Emil Robert von (1822-1880)	**Hanst.**
Hara, Hiroshi (1911-1986)	**H.Hara**
Haračić, Ambrosio (1855-1916)	**Haračić**
Harley, Winifred Jewel (1895- ?)	**W.J.Harley**
Harling, Gunnar (Wilhelm) (1920-)	**Harling**
Harper, Roland McMillan (1878-1966)	**R.M.Harper**
Harriman, Edward Henry (1848-1909)	**E.H.Harriman**
Harrington, Mark Walrod (1848-1926)	**Harr.**
Harris, Stuart (Stewart) Kimball (1906-)	**S.K.Harris**
Harting, Pieter (1812-1885)	**Harting**
Hartman, Carl Johan(n) (1790-1849)	**Hartm.**
Hartman, Emily L. (1932-)	**E.L.Hartm.**
Hasskarl, Justus Carl (1811-1894)	**Hassk.**
Hassler, Émile (1861-1937)	**Hassl.**
Hatanaka, Shin-ichi (1932-)	**Hatan.**
Hatusima, Sumihiko (1906-)	**Hatus.**
Haufler, Christopher Hardin (1950-)	**Haufler**
Hauke, Richard Louis (1930-)	**Hauke**
Hausmann (Hausman), Franz von (1810-1878)	**Hausm.**
Hausmann zu Stetten, Franz von	
vide Hausmann, F. von	**Hausm.**
Haussknecht, Heinrich Carl (1838-1903)	**Hausskn.**
Hawkes, Alex Drum (1927-1977)	**A.D.Hawkes**
Hawkes, John Gregory (1915-1977)	**Hawkes**
Hayata, Bunzô (1874-1934)	**Hayata**
Hayek, August (Edler) von (1871-1928)	**Hayek**
Hazslin, Friedrich August (Frigyes Ágost)	
vide Hazslinszky von Hazslin, F.A.	**Hazsl.**
Hazslinszky, Friedrich August (Frigyes Ágost)	
vide Hazslinszky von Hazslin, F.A.	**Hazsl.**
Hazslinszky von Hazslin, Friedrich August (Frigyes	
Ágost) (1818-1896)	**Hazsl.**
He, Ji-Jian (1933-)	**J.J.He**
Hectot, Jean Alexandre (1764- ?)	**Hectot**
Hedwig, Johann (Joannis, Ioannis) (1730-1799)	**Hedw.**
Hedwig, Romanus Adolf (1772-1806)	**R.Hedw.**

Hegi, Gustav (1876-1932)	**Hegi**
Heimerl, Anton (1857-1942)	**Heimerl**
Heine, Hermann Heino (1922-)	**Heine**
Heinsen, Ernst (fl. 1894)	**Heinsen**
Heller, Amos Arthur (1867-1944)	**A.Heller**
Hellwig, Franz Carl (1861-1889)	**Hellw.**
Hemsley, William Botting (1843-1924)	**Hemsl.**
Henderson, Louis Forniquet (1853-1942)	**L.F.Hend.**
Hennedy, Roger (1809-1877)	**Hennedy**
Hennipman, Elbert (1937-)	**Hennipman**
Henriques, Julio Augusto (1838-1928)	**Henriq.**
Hergt, Bernhard Julius Eduard (1858-1920)	**B.Hergt**
Héribaud-Joseph (Héribaud, Joseph), Frère (né Caumel, Jean Baptiste) (1841-1918)	**Hérib.**
Héritier, Charles Louis L' *vide* L'Héritier de Brutelle, C.L.	**L'Hér.**
Hermann, Paul [latine Hermannus] (1646-1695)	**Herm.**
Herrero, Alberto (1968-)	**Herrero**
Herter, Wilhelm (Guillermo) Gustav(o) Franz (Francis) (1884-1958)	**Herter**
Hervier Basson, Gabriel Marie Joseph (1846-1900)	**Hervier**
Herzog, Theodor Carl (Karl) Julius (1880-1961)	**Herzog**
Hetterscheid, Wilbert Leonard Anna (1957-)	**Hett.**
Heufler, Ludwig Samuel Joseph David Alexander von (from 1865 Freiherr von Hohenbühel) (1817-1885)	**Heufl.**
Heufler zu Rasen und Perdonegg, Ludwig Samuel Joseph David Alexander von *vide* Heufler, L.S.J.D.A. von	**Heufl.**
Heukels, Hendrik (1854-1936)	**Heukels**
Heurck, Henri Ferdinand Van *vide* Van Heurck, H.F.	**Van Heurck**
Hevly, Richard Holmes (1934-)	**Hevly**
Heward, Robert (1791-1877)	**Heward**
Heywood, Vernon Hilton (1927-)	**Heywood**
Hicken, Cristóbal María (1875-1933)	**Hicken**
Hickey, Ralph James (1950-)	**Hickey**
Hickok, Leslie George (1946-)	**Hickok**
Hieronymus, Georg Hans Emmo (Emo) Wolfang (1846-1921)	**Hieron.**
Hiitonen, Henrik Ilmari Augustus (until 1932 Hidén) (1898-)	**Hiitonen**
Hill, John (1716-1775)	**Hill**
Hill, Walter (1820-1904)	**W.Hill**
Hillebrand, Wilhelm (William) (1821-1886)	**Hillebr.**
Hirmer, Max (1893-1981)	**Hirmer**
Hirth, Alfred (fl. 1908)	**Hirth**
Hitchcock, Albert Spear (né Jennings, A.S.) (1865-1935)	**Hitchc.**
Hitchcock, Edward (1793-1864)	**E.Hitchc.**

Hiyama, Kôzô (1905-)	**Hiyama**
Hobdy, Robert Warner (1942-)	**Hobdy**
Hochstetter, Christian Ferdinand Friedrich (1787-1860)	**Hochst.**
Hodge, Walter Hendricks (1912-)	**Hodge**
Hoehne, Frederico Carlos (1882-1959)	**Hoehne**
Hoehnel, Franz Xaver Rudolf von	
vide Höhnel, F.X.R. von	**Höhn.**
Hoek, L. Van	
vide Van Hoek, L.	**Van Hoek**
Hoffmann, Georg Franz (1760-1826)	**Hoffm.**
Hoffmann, Ralph (1870-1932)	**Ralph Hoffm.**
Hoffmannsegg, Johann Centurius von (1766-1849)	**Hoffmanns.**
Höfler, Karl (1893- ?)	**Höfler**
Hofmann, Hermann (? -1923)	**H.Hofm.**
Hohenacker, Rudolf Friedrich (1798-1874)	**Hohen.**
Hohenbühel-Heufler, Ludwig Samuel Joseph David Alexander von	
vide Heufler, L.S.J.D.A. von	**Heufl.**
Höhnel, Franz Xaver Rudolf von (1852-1920)	**Höhn.**
Holl, Friedrich (fl. 1830-1842)	**Holl**
Hollick, Charles Arthur (1857-1933)	**Hollick**
Holloway, John Ernest (1881-1945)	**Holloway**
Hollrung, Max Udo (1858-1937)	**Hollrung**
Holmberg, Otto Rudolf (1874-1930)	**Holmb.**
Holttum, Richard Eric (1895-1990)	**Holttum**
Holub, Joseph (1930-)	**Holub**
Holuby, Joseph (Jószef) Ludwig (Ludovit) (1836-1923)	**Holuby**
Hölzer, ? (fl. 1939-1943)	**Hölzer**
Homann, Georg Gothilf Jacob (fl. 1828-1835)	**Homann**
Hombron, Jacques Bernard (1800-1852)	**Hombr.**
Honda, Masaji (Masazi) (1897-1984)	**Honda**
Hooker, Joseph Dalton (1817-1911)	**Hook.f.**
Hooker, William Jackson (1785-1865)	**Hook.**
Hooper, Elisabeth Andrews (née Andrews, Elisabeth	
Gale) (1960-)	**E.A.Hooper**
Hoorebeke, Charles Joseph Van	
vide Van Hoorebeke, C.J.	**Van Hooreb.**
Hoover, Robert Francis (1913-1970)	**Hoover**
Hope, Charles William Webley (1832-1904)	**C.Hope**
Hopkins, Lewis Sylvester (1872-1945)	**Hopkins**
Hoppe, David Heinrich (1760-1846)	**Hoppe**
Horaninow (Horaninov), Paul (Paulus) Fedorowitsch (1796-1865)	**Horan.**
Hornemann, Jens Wilken (1770-1841)	**Hornem.**
Hornschuch, Christian Friedrich (1793-1850)	**Hornsch.**
Horsfield, Thomas (1773-1859)	**Horsf.**
Hoshizaki, Barbara (née Joe,B.) (1928-)	**Hoshiz.**
Hosokawa, Takahide (1909-)	**Hosok.**

Hosseus, Carl Curt (1878-1950)	**Hosseus**
Houlston, John (fl. 1848-1852)	**Houlston**
House, Homer Doliver (1878-1949)	**House**
Houtte, Louis Benoit Van	
vide Van Houtte, L.B.	**Van Houtte**
Houttuyn, Maarten (Martin, Martinus) (1720-1798)	**Houtt.**
Hovenkamp, Peter Hans (1953-)	**Hovenkamp**
Howard, Richard Alden (1917-)	**R.A.Howard**
Howe, Elliot Calvin (1828-1899)	**Howe**
Howe, W.E. (? -1891)	**W.E.Howe**
Howell, John Thomas (1903-)	**J.T.Howell**
Hsia, Chun	
vide Xia, Qun	**Q.Xia**
Hsieh, Chang-Fu (1947-)	**C.F.Hsieh**
Hsieh (Hsien), Yin-Tang (1924-)	**Y.T.Hsieh**
Hsing, Kung-Hsia	
vide Shing, Kung-Hsia	**K.H.Shing**
Hsu, Ying-Pen (Ying-Peng, Ying-Ben, Yang-Pen, Yang-Pong)	
(1933-)	**Y.P.Hsu**
Hu, Hsen-Hsu (1894-1968)	**Hu**
Huang, Chi-Chuang	
vide Wong, Kin-Kuang	**K.K.Wong**
Huang, Tseng-Chieng (1931-)	**T.C.Huang**
Hudson, William (1730-1793)	**Huds.**
Hughes, Ralph Harley (1913-)	**R.H.Hughes**
Hultén, Eric Oskar Gunnar (1894-1981)	**Hultén**
Humbert, Jean-Henri (1887-1967)	**Humbert**
Humboldt, Friedrich Wilhelm Heinrich Alexander	
von (1769-1859)	**Humb.**
Hunkin, Joseph Wellington (1887-1950)	**Hunkin**
Hunter, William (1755-1812)	**W.Hunter**
Huntington, John Warren (1853- ?)	**Huntington**
Husnot, Pierre Tranquille (1840-1929)	**Husn.**
Huter, Rupert (1834-1919)	**Huter**
Huth, Ernst (1845-1897)	**Huth**
Hutoh, Masakazu (1930-1972)	**Hutoh**
Huxley, Camilla Rosalind (1952-) ✻	**C.R.Huxley**
Hy, Félix Charles (1853-1918)	**Hy**
Hyde, Harold Augustus (1892-1973)	**Hyde**
Hylander, Nils (1904-1970)	**Hyl.**
Ikebe, Chikako (1953-)	**Ikebe**
Iljin, Modest Mikhailovich (1889-1967)	**Iljin**
im Thurn, Everard Ferdinand	
vide Thurn, E.F. im	**Thurn**
Iranzo Reig, Julio (1942-)	**Iranzo**

Irmscher, Edgard (1887-1968)	**Irmsch.**
Irudayaraj, Varaprasatham (1960-)	**Irud.**
Itô, Hiroshi (1909-)	**H.Itô**
Ivanenko, Yury Alexeevich (1962-)	**Ivanenko**
Iversen, Johannes (1904-1971)	**Iversen**
Ivery, James (c.1823-1872)	**Ivery**
Iwatsuki, Kunio (1934-)	**K.Iwats.**
Jackson, Charles (fl. 1850s-1860s)	**C.Jacks.**
Jackson, Peter John (1958-)	**P.J.Jacks.**
Jacob de Cordemoy, Eugène	
vide Cordemoy, E. Jacob de	**Cordem.**
Jacobsen, Niels Henning Gunther (1941-)	**N.Jacobsen**
Jacobsen, Werner Bahne Georg (1909-)	**W.Jacobsen**
Jacquemont, Venceslas Victor (1801-1832)	**Jacquem.**
Jacques-Félix, Henri (1907-)	**Jacq.-Fél.**
Jacquin, Nicolaus (Nikolaus, Nicolaas) Joseph von (1727-1817)	**Jacq.**
Jacquinot, Honoré (1814-1887)	**Jacquinot**
Jahandiez, Émile (1876-1938)	**Jahand.**
Jaman, Razali (fl. 1987-1989)	**Jaman**
James, J. (fl. 1850s)	**J.James bis**
Jamir, N.S. (fl. 1981-1989)	**Jamir**
Jan, Georg (Giorgio) (1791-1866)	**Jan**
Janchen, Erwin Emil Alfred (1882-1970)	**Janch.**
Janchen-Michel von Westland, Erwin Emil Alfred	
vide Janchen, E.E.A.	**Janch.**
Jansen, Johannes Theodorus (1890-1948)	**J.T.Jansen**
Jansen, Pieter (1882-1955)	**Jansen**
Jaquotot, María Concepción (1932-)	**Jaquotot**
Jaquotot Villalonga, María Concepción	
vide Jaquotot, M.C.	**Jaquotot**
Jarrett, Frances Mary (1931-)	**F.M.Jarrett**
Jarvis, Charlie (Edward) (1954-)	**C.E.Jarvis**
Jaubert, Hippolyte François (1798-1874)	**Jaub.**
Jeanpert, (Henry) Édouard (1861-1921)	**Jeanp.**
Jenman, Georg Samuel (1845-1902)	**Jenman**
Jennings, Otto Emery (1877-1964)	**Jenn.**
Jepson, Willis Linn (1867-1946)	**Jeps.**
Jermy, Anthony Clive (1932-)	**Jermy**
Jervis, Swinfen (fl. 1860s)	**S.Jervis**
Jessen, Stefan (1956-)	**S.Jess.**
Jessop, John Peter (1939-)	**Jessop**
Jewell, Herbert Winship (1872- ?)	**Jewell**
Jha, Jaykar (fl. 1988)	**J.Jha**
Jiménez, Oton (1895-)	**Jiménez**
Jin, Yue-Xing (1934-)	**Y.X.Jin**

Jing, Yue-Xing	
vide Jin, Yue-Xing	**Y.X.Jin**
Joaquín de Barnola, María	
vide Barnola, J.M. de	**Barnola**
Joe, Barbara (later Hoshizaki, B.) (1928-)	**Joe**
Joenckema, Rembert van	
vide Dodoens, R.	**Dodoens**
Johns, Robert James (1944-)	**R.J.Johns**
Johnson, David Mark (1955-)	**D.M.Johnson**
Johnson, Eileen Ruth Laithlain (1896-1972)	**E.R.L.Johnson**
Johnson, James Yate (1820-1900)	**J.Y.Johnson**
Johnson-Groh, Cindy Lee (1957-)	**Johnson-Groh**
Johnston, Harry Hamilton (1858-1927)	**H.H.Johnst.**
Johnston, Ivan Murray (1898-1960)	**I.M.Johnst.**
Johow, Friedrich (Federico) Richard Adalbert (Adelbart) (1859-1933)	**Johow**
Joncheere, Gerardus Johannes Pieter de	
vide de Joncheere, G.J.P.	**de Jonch.**
Jones, Arthur Mowbray (1820-1889)	**A.M.Jones**
Jones, David Lloyd (1944-)	**D.L.Jones**
Jones, George Neville (1903-1970)	**G.N.Jones**
Jones, Marcus Eugene (1852-1934)	**M.E.Jones**
Jones, S. (fl. 1861)	**S.Jones**
Jordan, (Claude Thomas) Alexis (1814-1897)	**Jord.**
Joseph, J. (fl. 1964-1979)	**J.Joseph**
Jôtani, Yukio (1904-)	**Jôtani**
Junge, Paul (1881-1919)	**Junge**
Junghuhn, (Friedrich) Franz Wilhelm (1809-1864)	**Jungh.**
Jussieu, Antoine Laurent de (1748-1836)	**Juss.**
Jussieu, Bernard de (1699-1777)	**B.Juss.**
Jyothi, Puthiya Veetil (1968-)	**Jyothi**
Kachroo, Prem Nath (1924-)	**Kachroo**
Kallio, Paavo Pauli (1914-)	**Kallio**
Kanehira, Ryôzô (1882-1948)	**Kaneh.**
Kanitz, August (Agoston, Agost) (1843-1896)	**Kanitz**
Karsten, George (Henry Hermann) (1863-1937)	**G.Karst.**
Karsten, (Gustav Karl Wilhelm) Hermann (1817-1908)	**H.Karst.**
Kartesz, John T. (fl. 1991)	**Kartesz**
Kato, Masahiro (1946-)	**M.Kato**
Kaulfuss, Georg Friedrich (1786-1830)	**Kaulf.**
Kaur, Surjit (1936-)	**S.Kaur**
Kaur, Swaru Jeet (fl. 1975)	**S.J.Kaur**
Kavanagh, Kathryn Patricia (1953-)	**K.P.Kavanagh**
Kawasaki, Tsugio (1929-)	**T.Kawas.**
Keddy, P.A. (1953-)	**Keddy**

Keller, Louis (1850-1915)	**L.Keller**
Keller, Robert (1854-1939)	**R.Keller**
Kellogg, Albert (1813-1887)	**Kellogg**
Kerner, Anton Joseph von (1831-1898)	**A.Kern.**
Kerner von Marilaun, Anton Joseph	
vide Kerner, A.J. von	**A.Kern.**
Kersten, Otto (1839-1900)	**Kerst.**
Keyes, Janet A. (fl. 1993) ✳	**Keyes**
Keyserling, (Andreëvich) Alexander Friedrich Michael	
Leberecht Arthur von (1815-1891)	**Keyserl.**
Khandelwal, Sharda (1949-)	**Khand.**
Khokhrjakov (Khokhryakov), Andrey Pavlovich (1935-)	**A.P.Khokhr.**
Kholia, Bhupendra Singh (1962-)	**Kholia**
Khullar, Surinder Pal (1941-)	**Khullar**
Kiaerskov (Kiaerskou), Hjalmar Frederik Christian	
(1835-1900)	**Kiaersk.**
Kickx, Jean (filius) (1803-1864)	**J.Kickx f.**
Killip, Ellsworth Paine (1890-1968)	**Killip**
Kinahan, John Robert (1828-1863)	**Kinahan**
Kinney, Abbot (1850-1920)	**Kinney**
Kinney, Clarence W. (fl. 1936)	**C.W.Kinney**
Kirk, Thomas (1828-1898)	**Kirk**
Kirouak, Joseph Louis Conrad (later Frère Marie Victorin)	
vide Victorin, Frère Marie	**Vict.**
Kirschleger, Frédéric (Friedrich) R. (1804-1869)	**Kirschl.**
Kiss, Árpád (1889-1968)	**Kiss**
Kiss von Zilah, Endre (1873- ?)	**E.Kiss**
Kitagawa, Masao (1909-)	**Kitag.**
Kitaibel, Pál (Paul) (1757-1817)	**Kit.**
Kittredge, Elsie May (1870-1954)	**Kittr.**
Kjellberg, Gunnar Kostantin (1885-1943)	**Kjellb.**
Kjellqvist, Ebbe (1931-1989)	**Kjellq.**
Klinge, Johannes Christoph (1851-1902)	**Klinge**
Klinggräff, Hugo Erich Meyer von (1820-1902)	**H.Klinggr.**
Klinsmann, Ernst Ferdinand (1794-1865)	**Klinsm.**
Klobukova, Eugenija Nikolaevna Alissova	
vide Alissova-Klobukova, E.N.	**Aliss.**
Klotzsch, Johann Friedrich (1805-1860)	**Klotzsch**
Knobloch, Irving William (1907-)	**Knobloch**
Knoll, Fritz (1883-1981)	**Knoll**
Knowlton, Frank Hall (1860-1926)	**Knowlt.**
Knox, Elizabeth M. (1899-)	**E.M.Knox**
Knuth, Reinhard Gustav Paul (1874-1957)	**R.Knuth**
Koch, Karl (Carl) Heinrich Emil (Ludwig) (1809-1879)	**K.Koch**
Koch, Walo (1896-1956)	**W.Koch**
Koch, Wilhelm Daniel Joseph (1771-1849)	**W.D.J.Koch**

Koch-Grünberg, Christian Theodor (1872- ?)	**Koch–Grünb.**
Kodama, Shinsuke (1884-1944)	**Kodama**
Koenig, Johann Gerhard	
vide König, J.G.	**J.König**
Koernicke, Friedrich August	
vide Körnicke, F.A.	**Körn.**
Köhler, Erich (1889- ?)	**Er.Köhler**
Koidzumi, Gen'ichi (Genichi, Geniti, Gen-Iti) (1883-1953)	**Koidz.**
Køie, Mogens Engell (1911-)	**Køie**
Kolhatkar, G.G. (1902-)	**Kolh.**
Komarov, Vladimir Leontjevich (Leontevich) (1869- 1945)	**Kom.**
Kong, Xian-Xu	
vide Kung, Hsian-Shiu	**H.S.Kung**
König, Johann Gerhard (1728-1785)	**J.König**
Kopp, Julius (1880- ?)	**Kopp**
Kornaś, Jan (1923-1994)	**Kornaś**
Körnicke, Friedrich August (1828-1908)	**Körn.**
Kossinsky, Constantin Constantinovich (1874-1923)	**Kossinsky**
Kott, Laima Solveiga (née Zichmanis, L.S.) (1946-)	**Kott**
Kotuchov, Yury Andreevich (fl. 1966)	**Kotuchov**
Koyama, Tetsuo Michael (1933-)	**T.Koyama**
Krajina, Vladimir (Joseph) (1905-1993)	**Krajina**
Kramer, Karl Ulrich (1928-1994)	**K.U.Kramer**
Krasnoborov, Ivan M. (1931-)	**Krasnob.**
Krasser, Fridolin (1863-1923)	**Krasser**
Kreczetowicz (Kreczetovicz, Kretschetovitsch), Vitaly	
Ivanovich (1901-1942)	**V.I.Krecz.**
Krieger, Walther (1880- ?)	**W.Krieg.**
Krishna, Bal (fl. 1990)	**Bal Krishna**
Krok, Thorgny Ossian Bolivar Napoleon (1834-1921)	**Krok**
Krug, Carl (Karl) Wilhelm Leopold (1833-1898)	**Krug**
Kubitzki, Klaus (1933-)	**Kubitzki**
Kudô, Yûshun (1887-1932)	**Kudô**
Kugler, E. (fl. 1874-1883)	**Kugler**
Kühlewein, Paul Eduard (1798-1870)	**Kühlew.**
Kuhn, (Friedrich Adalbert) Maximilian ("Max") (1842-1894)	**Kuhn**
Kukkonen, Ilkka (Toiva Kalervo) (1926-)	**Kukkonen**
Kümmerle, Jenö Belá (1876-1931)	**Kümmerle**
Kung, Hsian-Shiu (1930-)	**H.S.Kung**
Kunkel, Günther (1928-)	**G.Kunkel**
Kunth, Karl (Carl) Sigismund (1788-1850)	**Kunth**
Kuntze, Carl (Karl) Ernst (Eduard) Otto (1843-1907)	**Kuntze**
Kunze, Gustav (1793-1851)	**Kunze**
Kuo, Chen Meng (1948-)	**C.M.Kuo**
Kurata, Satoru (1922-1978)	**Sa.Kurata**
Kurita, Siro (1936-)	**Kurita**

Kurz, Wilhelm Sulpiz (later Amann, Johann) (1834-1878)	**Kurz**
Kusnezow (Kusnezov), Nicolai Ivanovitch (1864-1932)	**Kusn.**
Kuvaev, Vladimir Borisovich (1918-)	**Kuvaev**
Kuvajev, Wladimir Borisovich	
vide Kuvaev, V.B.	**Kuvaev**
Kuzeneva, Olga Iakinfovna (1887-1978)	**Kuzen.**
Kuzeneva-Prochkorova, Olga Iakinfovna	
vide Kuzeneva, O.I.	**Kuzen.**
Kuznetzov (Kuznetzou), Nicolai Ivanovitch	
vide Kusnezow, N.I.	**Kusn.**
L'Escluse (L'Ecluse), Charles (Carolus) de [latine Clusius]	
(1526-1609)	**Clus.**
L'Héritier de Brutelle, Charles-Louis (1746-1800)	**L'Hér.**
L'Herminier, Ferdinand (1802-1866)	**L'Herm.**
L'Obel, Matthias de	
vide Lobel, M. de	**Lobel**
La Billardière, Jacques Julien Houtton de	
vide Labillardière, J.J.H. de	**Labill.**
La Gasca, Mariano de	
vide Lagasca y Segura, M. de	**Lag.**
La Marck, Jean Baptiste Antoine Pierre de Monnet de	
vide Lamarck, J.B.A.P. de Monnet de	**Lam.**
la Peirouse, Philippe Picot de	
vide Lapeyrouse, P. Picot de	**Lapeyr.**
La Pylaie, Auguste Jean Marie Bachelot de	
vide Bachelot de la Pylaie, A.J.M.	**Bach.Pyl.**
Labillardière, Jacques Julien Houtton de (1755-1834)	**Labill.**
Lacaita, Charles Charmichael (1853-1933)	**Lacaita**
Lachmann, Jean Paul (1851-1907)	**P.Lachm.**
Lackström, Emil Frithiof (? -1883)	**Lackström**
Laestadius, Lars Levi (1800-1861)	**Laest.**
Lagasca y Segura, Mariano de (1776-1839)	**Lag.**
Laguna y Villanueva, Máximo (1826-1902)	**Laguna**
Lai, Ming-Jou (1949-)	**M.J.Lai**
Laine, Unto Olavi (1930-)	**Laine**
Laing, Robert Malcolm (1865-1941)	**Laing**
Laínz Gallo, Manuel	
vide Laínz, M.	**Laínz**
Laínz, Manuel (1923-)	**Laínz**
Lam, Herman Johannes (1892-1977)	**H.J.Lam**
Lamarck, Jean Baptiste Antoine Pierre, de Monnet de (1744-1829)	**Lam.**
Lan, Young-Chen	
vide Lan, Young-Zhen	**Y.Z.Lan**
Lan, Young-Zhen (Yong-Zhen) (1933-)	**Y.Z.Lan**
Landry, Garrie Paul (1951-)	**G.P.Landry**

Lane, Irwin E. (1926-)	**Lane**
Láng, Adolph (Adolf) Franz (1795-1863)	**Láng**
Lang, Frank Alexander (1937-)	**F.A.Lang**
Lange, Johan Martin Christian (1818-1898)	**Lange**
Langsdorff, Georg Heinrich von (1774-1852)	**Langsd.**
Lapeyrouse (Lapeirouse), Philippe, Picot de (1744-1818)	**Lapeyr.**
Large, Mark Frederick (1959-)	**Large**
Larrañaga, Dámaso Antonio (1771-1848)	**Larrañaga**
Lasch, Wilhelm Gottfried (1787-1863)	**Lasch**
Lasser, Tobias (1911-)	**Lasser**
Laubenburg, Karl Ernst (fl. 1899)	**Laubenb.**
Lauche, (Friedrich) Wilhelm (Georg) (1827-1883)	**Lauche**
Launert, Oskar Edmund (1926-)	**Launert**
Lauterbach, Carl (Karl) Adolf Georg (1864-1937)	**Lauterb.**
Lawalrée, André Gilles Célestin (1921-)	**Lawalrée**
Lawson, George (1827-1895)	**G.Lawson**
Lazarev, Anatoly Michilovich (fl. 1981)	**Lazarev**
Le Grand, Antoine	
vide Legrand, A.	**Legrand**
Lecomte, Paul Henri (1856-1934)	**Lecomte**
Ledebour, Carl (Karl) Friedrich von (1785-1851)	**Ledeb.**
Lee, Alma Theodora (née Melvaine, A.T.) (1912-1990)	**A.T.Lee**
Lee, Chang-Shook (1957-) *	**C.S.Lee**
Leeds, Arthur Newlin (1870-1939)	**Leeds**
Legrand, Antoine (1839-1905)	**Legrand**
Legrand, Carlos Maria Diego Enrique (1901-)	**D.Legrand**
Lehmann, Johann Georg Christian (1792-1860)	**Lehm.**
Leisman, Gilbert Arthur (1924-)	**Leisman**
Lejeune, Alexandre (Alexander) Louis (Ludwig) Simon (1779-1858)	**Lej.**
Lellinger, David Bruce (1937-)	**Lellinger**
Lélorgne de Savigny, (Marie) Jules César	
vide Savigny, (M.)J.C. Lélorgne de	**Savigny**
Lemaire, (Antoine) Charles (1800-1871)	**Lem.**
Léman, Dominique Sébastien (1781-1829)	**Léman**
Lemmon, John Gill (1832-1908)	**Lemmon**
Léon, Blanca (1957-)	**B.Léon**
Léon, Frère (né Sauget y Barbis, Joseph Sylvestre) (1871-1955)	**Léon**
Lepage, Ernest (1905-1981)	**Lepage**
Leprieur, F.M.R. (1799-1869)	**Lepr.**
Leresche, Louis François Jules Rodolphe (1808-1885)	**Leresche**
Lestiboudois, Thémistocle Gaspard (1797-1876)	**T.Lestib.**
Letourneux, Aristide-Horace (1820-1890)	**Letourn.**
Leunis, Johannis (1802-1873)	**Leunis**
Léveillé, Augustin Abel Hector (1863-1918)	**H.Lév.**
Levier, Émile (Emilio) (1839-1911)	**Levier**

Levin, Geoffrey Arthur (1955-)	**G.A.Levin**
Levinge, Harry Corbyn (1831-1896)	**Levinge**
Leybold, Friedrich (1827-1879)	**Leyb.**
Li, Fa-Zeng (1942-)	**F.Z.Li**
Li, Feng (1957-)	**F.Li**
Li, Hua (1962-)	**Hua Li**
Li, Hui-Lin (1911-) ✱	**H.L.Li**
Li, Jian-Xiu (1937-)	**J.X.Li**
Li, Na (1965-)	**N.Li**
Li, Shu-Xin (Shu-Hsin, Chu-Xin) (1926-)	**S.X.Li**
Lichtenstein, Juana (née de Schafer, J.) (1902-)	**J.S.Licht.**
Lid, Johannes (1886-1971)	**Lid**
Liebmann, Frederik Michael (1813-1856)	**Liebm.**
Lightfoot, John (1735-1788)	**Lightf.**
Liljeblad, Samuel (1761-1815)	**Lilj.**
Lin, Su-Juan (fl. 1994)	**S.J.Lin**
Lin, You-Xing (You-Xin) (1934-)	**Y.X.Lin**
Lindau, Gustav (1866-1923)	**Lindau**
Linden, Jean Jules (1817-1898)	**Linden**
Lindig, Alejandro (Alexander) (fl. 1861)	**Lindig**
Lindley, John (1799-1865)	**Lindl.**
Lindman, Carl Axel Magnus (1856-1928)	**Lindm.**
Lindsay, William Lauder (1829-1880)	**Linds.**
Ling, You-Xin	
vide Lin, You-Xing	**Y.X.Lin**
Link, Johann Heinrich Friedrich (1767-1851)	**Link**
Linnaeus, Carolus (Carl) (1707-1778)	**L.**
Linné, Carl von (pater)	
vide Linnaeus, C.	**L.**
Linné, Carl von (filius) (1741-1783)	**L.f.**
Linton, William James (1812- 1898)	**Linton**
Liogier Allut, Enrique Eugenio (later Hermano Alain)	
vide Alain, Hermano	**Alain**
Liogier, Enrique Eugenio (later Hermano Alain)	
vide Alain, Hermano	**Alain**
Liogier, Henri Alain (later Hermano Alain)	
vide Alain, Hermano	**Alain**
Liou, Ming (1970-) ✱	**M.Liou**
Liou, Tchen-Ngo (1897-1975)	**T.N.Liou**
Litardière, René Verriet de (1888-1957)	**Litard.**
Liu, Zheng-Yu (1952-)	**Z.Y.Liu**
Lloyd, Francis Ernest (1868-1947)	**F.E.Lloyd**
Lloyd, James (1810-1896)	**J.Lloyd**
Lloyd, Robert Michael (1938-1994) ✱	**R.M.Lloyd**
Lobb, William (1809-1863)	**W.Lobb**
Lobel, Matthias de [latine Lobelius] (1538-1616)	**Lobel**

Lobin, Wolfram (1951-)	**Lobin**
Loidi Aguirre, Javier José	
vide Loidi, J.J.	**Loidi**
Loidi, Javier José (1952-)	**Loidi**
Loiseleur des Longchamps, Jean Louis Auguste	
vide Loiseleur-Deslongchamps, J.L.A.	**Loisel.**
Loiseleur-Deslongchamps, Jean Louis August(e)	
(1774-1849)	**Loisel.**
Loiacono-Poiero, Michele	
vide Lojacono Pojero, M.	**Lojac.**
Lojacono, Michele	
vide Lojacono Pojero, M.	**Lojac.**
Lojacono Pojero, Michele (1853-1919)	**Lojac.**
Lombardo, Atilio (fl. 1958)	**Lombardo**
Long, Gilbert August (1928-)	**G.Long**
Longa, Massimo (1854-1928)	**Longa**
Looser, Gualterio (1898-1982)	**Looser**
Lorea-Hernández, Francisco Gerardo (1956-)	**Lorea-Hern.**
Lorence, David Harold (1946-)	**Lorence**
Losch, Ilse (1920-)	**I.Losch**
Loss, Giuseppe (Josef) (1831-1880)	**Loss**
Lotsy, Johannes Paulus (1867-1931)	**Lotsy**
Loudon, John Claudius (1783-1843)	**Loudon**
Loureiro, João de (1717-1791)	**Lour.**
Lourteig, Alicia (1913-)	**Lourteig**
Löve, Åskell (1916-1994)	**Å.Löve**
Löve, Doris Benta Maria (1918-)	**D.Löve**
Lovis, John Donald (1930-)	**Lovis**
Lowe, Edward Joseph (1825-1900)	**E.J.Lowe**
Lowe, Richard Thomas (1802-1874)	**Lowe**
Loyal, Dasonda Singh (1926-1992)	**Loyal**
Lu, Shu-Gang (1957-)	**S.G.Lu**
Ludwig, Christian Gottlieb (1709-1773)	**Ludw.**
Luebke, Neil Thomas (1950-)	**Luebke**
Luerssen, Christian (1843-1916)	**Luerss.**
Lumpkin, T.A. (fl. 1989) *	**Lumpkin**
Lunell, Joël (1851-1920)	**Lunell**
Lusina, Giuseppe (1893-1963)	**Lusina**
Lyell, Katherine Murray (née Horner, K.M.) (1817-1915)	**K.Lyell**
Lyon, Harold Lloyd (1879-1957)	**Lyon**
M'Ken, Mark John (Johnston)	
vide McKen, M.J.	**McKen**
Ma, Chi-Yun (1942-)	**C.Y.Ma**
Ma, Shu-Tai (1955-)	**S.T.Ma**
Ma, Yi-Lun (fl. 1985-1987)	**Y.L.Ma**

Maack (Maak), Richard Karlovich (1825-1886)	**Maack**
Mabberley, David John (1948-)	**Mabb.**
Macbride, James Francis (1892-1976)	**J.F.Macbr.**
Machado Bueno, Rogerio	
vide Bueno, R. Machado	**R.M.Bueno**
Mackay, Alexander Howard (1848-1929)	**A.Mackay**
Macken, Mark John	
vide McKen, M.J.	**McKen**
Mackenzie, Kennet Kent (1877-1934)	**Mack.**
Macloskie, George (1834-1919)	**Macloskie**
Madhusoodanan (Madhusoodanam), Pandara Valappil (1950-)	**Madhus.**
Maehara, Kanjiro	
vide Mayebara, K.	**Mayeb.**
Maekawa, Fumio (1908-1984)	**F.Maek.**
Magnier, Charles (fl. 1884)	**Magnier**
Magnol, Pierre (1638-1715)	**Magnol**
Maguire, Bassett (1904-1991)	**Maguire**
Mahabalé, Tryambak Shankur (1909-1983)	**Mahab.**
Maiden, Joseph Henry (1859-1925)	**Maiden**
Maillard, Pierre Néhémie (1813-1883)	**Maillard**
Maire, René Charles Joseph Ernest (1878-1949)	**Maire**
Mäkinen, Yrjö Lauri Antero (1931-)	**Y.Mäkinen**
Makino, Tomitarô (1862-1957)	**Makino**
Malik, Siddiqa (1940-)	**S.Malik**
Malinvaud, Louis Jules Ernest (1836-1913)	**Malinv.**
Malla, Samar Bahadur (1933-)	**S.B.Malla**
Malý, Karl (Franz Josef) (1874-1951)	**K.Malý**
Manetti, Saverio (Xaverius) (1723-1785)	**Manetti**
Manickam, Visuvasam Soosai (1944-)	**Manickam**
Mann, Horace (1844-1868)	**H.Mann**
Manton, Irene (1904-1988)	**Manton**
Maratti, Giovanni Francesco (Johannes Franciscus) (1723-1777)	**Maratti**
Marcano, Vicente (fl. 1989)	**V.Marcano**
Marchesetti, Carlo de (Carl von) (1850-1926)	**Marches.**
Marchetti, Dino (1942-)	**Marchetti**
Marcucci, Emilio (1837-1890)	**Marcucci**
Marie-Victorin, Frère (né Kirouak, Joseph Louis Conrad)	
vide Victorin, Frère Marie	**Vict.**
Markgraf, Friedrich (1897-1987)	**Markgr.**
Márquez-Moya, Ana Luz (1966-)	**Márq.-Moya**
Marschall von Bieberstein, Friedrich August (1768-1826)	**M.Bieb.**
Marschlins, Carl Ulisses Adalbert von Salis-	
vide Salis-Marschlins, C.U.A. von	**Salis**
Marsden, Craigh Russell (1952-)	**C.R.Marsden**
Martens, Martin (1797-1863)	**M.Martens**
Marthe, François (fl. 1801)	**Marthe**

Martius, Carl (Karl) Friedrich Philipp von (1794-1868)	**Mart.**
Masalles, Ramón Maria (1948-)	**Masalles**
Masamune, Genkei (1899-)	**Masam.**
Massalongo, Caro Benigno (1852-1928)	**C.Massal.**
Masters, Maxwell Tylden (1833-1907)	**Mast.**
Matsuda, Sadahisa (1857-1921)	**Matsuda**
Matsumoto, Sadamu (1947-)	**S.Matsumoto**
Matsumura, Jinzô (1856-1928)	**Matsum.**
Matthew, Charles Geekie (1862-1936)	**C.G.Matthew**
Matthews, James Francis (1935-)	**J.F.Matthews**
Mattioli, Pietro Andrea (Petrus Andrea, Pier Andrea) *[latine Matthiolus]* (1500-1577)	**Mattioli**
Mattirolo, Oreste (1856-1947)	**Mattir.**
Matuda, Eizi (1894-1978)	**Matuda**
Maury, Paul Jean Baptiste (1858-1893)	**Maury**
Maximowicz, Carl Johann (Ivanovič) (1827-1891)	**Maxim.**
Maxon, William Ralph (1877-1948)	**Maxon**
Maxwell, T.C. (1822-1908)	**Maxwell**
Maycock, Paul F. (fl. 1963)	**P.F.Maycock**
Mayebara, Kanjiro (1890- ?)	**Mayeb.**
Mayer, E. (fl. 1868)	**E.Mayer bis**
McAlpin, Bruce W. (1944-)	**McAlpin**
McColl, W.R. (1855-1933)	**McColl**
McCord, David Ross (1844- ?)	**McCord**
McCoy, Thomas Nevil (1905-)	**T.N.McCoy**
McDonald, William H. (1837-1902)	**McDonald**
McGregor, Ronald Leighton (1919-)	**McGregor**
McKen, Mark Johnston (John) (1823-1872)	**McKen**
McNab, William Ramsay (1844-1889)	**W.R.McNab**
McVaugh, Rogers (1909-)	**McVaugh**
Mehra, Pran Nath (1907-1994)	**Mehra**
Mehroff, Leslie J. (1950-)	**Mehroff**
Mehrotra, Bishan Narain (1936-)	**Mehrotra**
Meisner, Carl (Daniel) Friedrich (1800-1874)	**Meisn.**
Meissner, Carl (Daniel) Friedrich *vide* Meisner, C. (D.).F.	**Meisn.**
Melvaine, Alma Theodora (later Lee, A.T.) (1912-1990)	**Melvaine**
Melzer, Helmut (1922-)	**H.Melzer**
Mendenhall, John Mark (1959-)	**Mendenh.**
Mendes, Eduardo José (Santos Moreira) (1924-)	**Mendes**
Mendonça, Francisco de Ascensão (1889-1982)	**Mendonça**
Menezes (Meneses), Carlos Azevedo de (1863-1928)	**Menezes**
Mentzel, Christian [latine Mentzelius] (1622-1701)	**Mentzel**
Menzies, Archibald (1754-1842)	**Menzies**
Mer, Émile (1841-1921)	**Mer**
Mérat de Vaumartoise, François Victor (1780-1851)	**Mérat**

Merino, Baltasar (1845-1917) **Merino**
Merino y Román, (Padre) Baltasar
   vide Merino, B. **Merino**
Merrill, Elmer Drew (1876-1956) **Merr.**
Mertens, Franz Karl (Carl) (1764-1831) **Mert.**
Mesa, Aldo (fl. 1971) **Mesa**
Mettenius, Georg Heinrich (1823-1866) **Mett.**
Meyen, Franz Julius Ferdinand (1804-1840) **Meyen**
Meyer, Carl (Karl) Anton (Andreevič, Andrejewicz) von (1795-1855) **C.A.Mey.**
Meyer, Dieter Erich (1926-1982) **D.E.Mey.**
Meyer, Georg Friedrich Wilhelm (1782-1856) **G.Mey.**
Meyer, Johann Carl Friedrich (1739-1811) **J.C.F.Mey.**
Meyer, Miriam Hope (née Wysong, M.H.) (1940-) **M.Mey.**
Miao, Ru-Huai
   vide Miau, Ru-Huai **R.H.Miau**
Miau, Ru-Huai (Ru-Hwai) (1942-) **R.H.Miau**
Michaux, André (1746-1802/3) **Michx.**
Micheli, Marc (1844-1902) **Micheli**
Micheli, Pier'Antonio (Pietro Antonio) [latine Michelius]
   (1679-1737) **P.Micheli**
Mickel, John Thomas (1934-) **Mickel**
Miègeville, Joseph (1819-1901) ✱ **Miègev.**
Miers, John (1789-1879) **Miers**
Mikheladze, Irakly Alexandrovich (?1937-) **Mikhel.**
Mildbraed, Gottfried Wilhelm Johannes (1879-1954) **Mildbr.**
Milde, (Carl August) Julius (1824-1871) **Milde**
Mill, Robert Reid (1950-) **R.R.Mill**
Mille, Luis (Louis Aloysius) (1873-1953) **Mille**
Miller, Elihu Sanford (1848-1940) **E.S.Mill.**
Miller, Harvey Alfred (1928-) **H.A.Mill.**
Millspaugh, Charles Frederick (1854-1923) **Millsp.**
Minamitani, Tadashi (1940-) **Minamit.**
Miniaev, Nicolai Aleksandrovich (1909-) **Miniaev**
Miquel, Friedrich Anton Wilhelm (1811-1871) **Miq.**
Mirbel, Charles François Brisseau de (1776-1854) **Mirb.**
Mitchell, David Searle (1935-) **D.S.Mitch.**
Mitchell, John (1690 or 1711-1768) **Mitch.**
Mitsuta, Shigeyuki (1950-) **Mitsuta**
Mitui, Kunio (1940-1988) **Mitui**
Miyabe, Kingo (1860-1951) **Miyabe**
Miyabe, Tsutomu (Tsutome) (1880-1921) **T.Miyabe**
Miyake, Kiichi (1876-1964) **Miyake**
Miyamoto, Futoshi (1958-) **Miyam.**
Miyazaki, Taku (1965-) **Miyazaki**
Mizushima, Masami (1925-1972) **M.Mizush.**
Mo, Sin-Li (1934-1988) **S.L.Mo**

Mo, Xin-Li	
vide Mo, Sin-Li	**S.L.Mo**
Moçiño (Mociño, Mocigno), José Mariano (1757-1820)	**Moç.**
Moçiño Suares de Figueroa, José Mariano	
vide Moçiño, J.M.	**Moç.**
Moçiño Suares Losada, José Mariano	
vide Moçiño, J.M.	**Moç.**
Moench, Conrad (1744-1805)	**Moench**
Mohlenbrock, Robert (H.) (1931-)	**Mohlenbr.**
Mohr, Daniel Matthias Heinrich (1780-1808)	**D.Mohr**
Mokry, Franz (1925-1986)	**Mokry**
Molesworth, Betty Eleanor Gosset Allen	
vide Allen, B.E.G. Molesworth	**B.M.Allen**
Molina, Giovanni Ignazio (né Juan Ignacio) (1740-1829)	**Molina**
Møller, Sophie Maren (1840-1920)	**S.M.Møller**
Molly, Mulanjanickal Joseph (1954-)	**Molly**
Momose, Sizuo (1906-1968)	**Momose**
Mondal, Papia (later Roy Choudhury, P.) (1957-)	**P.Mondal**
Monkman, Charles (fl. 1860s)	**Monkman**
Monnet (Monet) de Lamarck, Jean Baptiste Antoine Pierre de	
vide Lamarck, J.B.A.P. de Monnet de	**Lam.**
Monnier, Paul Camille Jacques (1922-)	**P.Monnier**
Montgomery, James Douglas (1937-)	**J.D.Montgom.**
Montserrat, Pedro	
vide Montserrat Recoder, P.	**P.Monts.**
Montserrat Recoder, Pedro (1918-)	**P.Monts.**
Moore, Albert Hanford (1883- ?)	**A.H.Moore**
Moore (Moir), David (1808-1879)	**Moore**
Moore, John William (1901-)	**J.W.Moore**
Moore, Lucy Beatrice (1906-1987)	**L.B.Moore**
Moore, Thomas (1821-1887)	**T.Moore**
Mora-Osejo, Luis Eduardo (1931-)	**L.E.Mora**
Moran, Reid Venable (1916-)	**Moran**
Moran, Robbin Craig (1956-)	**R.C.Moran**
Moreno Senna, Rosana	
vide Senna, R. Moreno	**R.M.Senna**
Mori, Tamezô (1884-1962)	**T.Mori**
Morison, Robert (1620-1683)	**Morison**
Moritz, Johann Wilhelm Karl (1797-1866)	**Moritz**
Moritzi, Alexander (1806-1850)	**Moritzi**
Morse, Warner Jackson (1872-1931)	**W.J.Morse**
Morton, Conrad Vernon (1905-1972)	**C.V.Morton**
Morton, Friedrich von (1890-1969)	**F.Morton**
Motelay, Léonce (1830-1917)	**Motelay**
Mougeot, Jean Baptiste (1776-1858)	**Moug.**
Moxley, George Loucks (1871- ?)	**Moxley**

Moziño Suarez de Figueroa, José Mariano
   vide Moçiño, J.M.   **Moç.**
Muehlenberg, Gotthilf Henry Ernest
   vide Muhlenberg, G.H.E.   **Muhl.**
Muehlenberg, Heinrich Ludwig
   vide Mühlenberg, H.L.   **H.L.Mühl.**
Mueller [Argoviensis], Johannes (Jean)
   vide Müller [Argoviensis], J.   **Müll.Arg.**
Mueller [Berolinensis], Karl (Carl)
   vide Müller [Berolinensis], K.(C.)   **Müll.Berol.**
Mueller, Ferdinand Jacob Heinrich von (1825-1896)   **F.Muell.**
Mueller [Halensis], Johann Karl (Carl) August (Friedrich Wilhelm)
   vide Müller [Halensis], J.K.(C.) A. (F.W.)   **Müll.Hal.**
Mueller, Otto Friedrich (Friderich, Fridrich, Frederik)
   vide Müller, O.F.   **O.F.Müll.**
Muhlenberg, Gotthilf Henry Ernst (1753-1815)   **Muhl.**
Mühlenberg, Henrich Ludwig (1756-1817)   **H.L.Mühl.**
Muir, David
   vide Moore, D.   **Moore**
Mules, F. (fl. 1860s)   **Mules**
Müller [Argoviensis], Johannes (Jean) (1828-1896)   **Müll.Arg.**
Müller [Berolinensis], Karl (Carl) (1817-1870)   **Müll.Berol.**
Müller, Ferdinand Jacob Heinrich von
   vide Mueller, F.J.H. von   **F.Muell.**
Müller [Halensis], Johann Karl (Carl) August (Friedrich
  Wilhelm) (1818-1899)   **Müll.Hal.**
Müller, Otto Friedrich (Friderich, Fridrich, Frederik) (1730-1784)   **O.F.Müll.**
Müller-Knatz, Johann Jacob (fl. 1912)   **Müll.-Knatz**
Mulligan, Gerald Alfred (1928-   )   **G.A.Mulligan**
Münderlein, Pfarrer (fl. 1898)   **Münderl.**
Muñoz Garmendia, José Félix (1949-   )   **Muñoz Garm**.
Munz, Philip Alexander (1892-1974)   **Munz**
Murakami, Noriaki (1959-   )   **N.Murak.**
Murbeck, Svante Samuel (1859-1946)   **Murb.**
Murchinson, Roderick Impey (1792-1871)   **Murch.**
Murdy, William Henry (1928-   )   **Murdy**
Murillo, María Teresa (1929-   )   **M.T.Murillo**
Murillo Pulido, María Teresa
   vide Murillo, M.T.   **M.T.Murillo**
Murray, Johan Andreas (Anders) (1740-1791)   **Murray**
Murto, Risto (1955-   )   **Murto**
Mus Amézquita, Maurici
   vide Mus, M.   **Mus**
Mus, Maurici (1961-   )   **Mus**
Mussche, Jean Henri (1765-1834)   **Mussche**
Mutel, Pierre Auguste Victor (1795-1847)   **Mutel**

Nábělek, František (1884-1965)	**Nábělek**
Nadeaud, Jean (1834-1898)	**Nadeaud**
Nag, Kalpana (1937-)	**Nag**
Nair, G. Bhadran (1942-)	**G.B.Nair**
Nair, Narayana (Pillai) Chandrashekharan (1927-)	**N.C.Nair**
Nakai, Takenoshin (Takenosin) (1882-1952)	**Nakai**
Nakaike, Toshiyuki (1943-)	**Nakaike**
Nakamura, Takehisa (1932-)	**T.Nakam.**
Nakashima, Kazuo (1904-1953)	**Nakash.**
Nakato, Narumi (1949-)	**Nakato**
Namegata, Tomitaro (1896-1970)	**Nameg.**
Nampy, Santhosh (1966-)	**Nampy**
Nardi, Enio (1942-)	**E.Nardi**
Nauman, Clifton Edward (1954-)	**Nauman**
Nayar, Bala Krishnan (1927-)	**B.K.Nayar**
Naylor, Frederick (1811-1882)	**Naylor**
Nazor, Vladimir (fl. 1903-1904)	**Nazor**
Necker, Noel (Martin) Joseph de (1730-1793)	**Neck.**
Nees, Christian Gottfried Daniel von Esenbeck	
vide Nees von Esenbeck, C.G.D.	**Nees**
Nees von Esenbeck, Christian Gottfried Daniel	
(1776-1858)	**Nees**
Neidorf, Charles (1915-)	**Neidorf**
Neilreich, August (1803-1871)	**Neilr.**
Nelson, Aven (1859-1952)	**A.Nelson**
Nelson, Elias Emanuel (1876-1949)	**E.E.Nelson**
Nemoto, Kwanji (1860-1936)	**Nemoto**
Nesom, Guy Lane (1945-)	**G.L.Nesom**
Nessel, Hermann (1877-1949)	**Nessel**
Newman, Edward (1801-1876)	**Newman**
Ni, Chen-Kai (1939-)	**C.K.Ni**
Nicholson, George (1847-1908)	**G.Nicholson**
Nicklès, Maurice (1907-)	**M.Nicklès**
Nicolson, Dan Henry (1933-)	**Nicolson**
Nicotra, Leopoldo (1846-1940)	**Nicotra**
Nieschalk, Albert (1904-1985)	**A.Niesch.**
Niessl von Mayendorf, Gustav (1839-1919)	**Niessl**
Nieto, Maria Lucila (1960-)	**Nieto**
Nieto-Caldera, José Maria (1955-)	**Nieto-Cald.**
Nieuwland, Julius (Aloysius) Arthur (1878-1936)	**Nieuwl.**
Nikolič, Emanuel (fl. 1904)	**Nikolič**
Nikolič, Vojislav (1925-1989)	**V.Nikolič**
Ninot, Josep Maria (1955-)	**Ninot**
Ninot i Sugrañes, Josep Maria	
vide Ninot, J.M.	**Ninot**

Nishida, Harufumi (1954-)	**H.Nishida**
Nishida, Makoto (1927-)	**M.Nishida**
Nishida, Toji (1874-1927)	**Nishida**
Nisman, Carmen (fl. 1971)	**Nisman**
Nisman S., Carmen	
vide Nisman, C.	**Nisman**
Nooteboom, Hans Peter (1934-)	**Noot.**
Noroña (Noronha), Francisco (François) (ca.1748-1787/88)	**Noronha**
Notaris, Giuseppe (Josephus) De	
vide De Notaris, G.(J.)	**De Not.**
Novák, František Antonín (1892-1964)	**Novák**
Nuttal, Thomas (1786-1859)	**Nutt.**
Nylander, Fredrik (Frederick) (1820-1880)	**F.Nyl.**
Nyman, Carl Fredrik (1820-1893)	**Nyman**
Oakes, William (1799-1848)	**Oakes**
Oberholzer, Ernst (fl. 1886-1950)	**Oberh.**
Obermeyer (Obermeijer, Obermejer), Anna Amelia	
(later Mauve, A.A.) (1907-)	**Oberm.**
Obermeyer-Mauve, Anna Amelia	
vide Obermeyer, A.A.	**Oberm.**
Oeder, George (Georg) Christian (from 1788 von Oldenburg)	
(1728-1791)	**Oeder**
Ogata, Masasuke (1883-1944)	**Ogata**
Ogle, Colin Charles (1941-)	**Ogle**
Ogura, Yudzuru (Yuzuro) (1895-1981)	**Ogura**
Oh, Yong-Cha (1940-) ✳	**Y.C.Oh**
Ohba, Hideaki (1943-)	**H.Ohba**
Ohmura, Toshiro (1920-)	**Ohmura**
Ohtani, Shigeru (1900-1981)	**Ohtani**
Ohwi, Jisaburo (1905-1977)	**Ohwi**
Oka, Kunio (1918-)	**Oka**
Oka, Taketoshi (1944-)	**T.Oka**
Oken, Lorenz (until 1802 Okenfuss) (1779-1851)	**Oken**
Okuyama, Shunki (1909-)	**Okuyama**
Oliver, Daniel (1830-1916)	**Oliv.**
Oliver, Walter Reginald Brook (1883-1957)	**W.R.B.Oliv.**
Øllgaard, Benjamin (1943-)	**B.Øllg.**
Olsson, Peter (1833-1906)	**P.Olsson**
Ooststroom, Simon Jan van (1906-1982)	**Ooststr.**
Opiz, Philipp Maximilian (1787-1858)	**Opiz**
Orbigny, (Alcide) Charles Victor Dessalines d' (1806- 1876)	**Orb.**
Orell Casasnovas, Jeroni (Jerónimo)	
vide Orell, J.	**Orell**
Orell, Jeroni (Jerónimo) (1924-)	**Orell**
Ormonde, José Eduardo Martins (1943-)	**Ormonde**

Ortega, Casimiro Gómez (de)
    vide Gómez (de) Ortega, C.   **Ortega**
Osten, Cornelius (1863-1936)   **Osten**
Otani, Yoshio (1919-   )   **Y.Otani**
Otsuka, Koichi (1951-   )   **Otsuka**

P'ang, Hsin-Min
    vide Pong, Sin-Min   **Pong**
Pacheco Mota, Armida Leticia (1956-   )   **Pacheco**
Pacher, David (1816-1902)   **Pacher**
Pacyna, Anna (1940-   )   **Pacyna**
Padley, Charles (fl. 1860s)   **Padley**
Paes do Amaral Franco, João Manuel António
    vide Franco, J.M.A. Paes do Amaral   **Franco**
Page, Christopher Nigel (1942-   )   **C.N.Page**
Paiva, Jorge Américo Rodriguez (1933-   )   **Paiva**
Pajarón, Santiago (1954-   )   **Pajarón**
Pajarón-Sotomajor, Santiago
    vide Pajarón, S.   **Pajarón**
Pal, Niranjan (1927-   )   **N.Pal**
Pal, Sunanda (1933-   )   **S.Pal**
Palacios-Rios, Mónica (1961-   )   **M.Palacios**
Paler, Michael H. (fl. 1995)   **Paler**
Palibin, Ivan Vladimirovich (1872-1949)   **Palib.**
Palisot de Beauvois, Ambroise Marie François Joseph (1752-1820)   **P.Beauv.**
Pallas, Peter (Pyotr) Simon (1741-1811)   **Pall.**
Palmer, Daniel Dooley (1930-   )   **D.D.Palmer**
Palmer, Thomas Chalkley (1860-1934)   **T.C.Palmer**
Palmer, William (1856-1921)   **W.Palmer**
Pampanini, Renato (1875-1949)   **Pamp.**
Pang, Sin-Min
    vide Pong, Sin-Min   **Pong**
Pangtey, Yash Pal Singh (1944-   )   **Pangtey**
Pangua, Emilia (1955-   )   **Pangua**
Panigrahi, Gopinath (1924-   )   **Panigrahi**
Pant, Divya Darshan (1918-   )   **D.D.Pant**
Panzer, Georg Wolfang Franz (1755-1829)   **Panz.**
Paoletti, Giulio (1865-1941)   **Paol.**
Papp, Constantin (1896-1972)   **Papp**
Pappe, Karl (Carl) Wilhelm Ludwig (1803-1862)   **Pappe**
Pardo, Cristina (1952-   )   **C.Pardo**
Paris, Cathy Ann (1953-   )   **C.A.Paris**
Parish, Samuel Bonsall (1838-1928)   **Parish**
Parker, Charles Sandbach (? -1869)   **C.Parker**
Parmentier, Antoine Auguste (1737-1813)   **Parm.**
Parmentier, Paul Evariste (1860-1946/47)   **P.Parm.**

Parris, Barbara Sydney (1945-)	**Parris**
Parrott, Fennella Jane Elizabeth (1962-) ✳	**F.J.E.Parrott**
Pascher, Adolf (Adolph) (1881-1945)	**Pascher**
Passerini, Giovanni (1816- 1893)	**Pass.**
Pastore, Ada Italia (1906-1952)	**Pastore**
Paterson, Robert H. (1814-1889)	**R.H.Paterson**
Patnaik, Satya Narain (Narayan) (1934-)	**Patnaik**
Pau, Carlos (1857-1937)	**Pau**
Pau y Español, Carlos	
vide Pau, C.	**Pau**
Paul, Alison Mary (1956-)	**A.M.Paul**
Paula von Schrank, Franz von	
vide Schrank, F. von Paula von	**Schrank**
Paulin, Alphons (1853-1942)	**Paulin**
Pavlov, Nikolai Vasilievich (1893-1971)	**Pavlov**
Pavón, José Antonio (1754-1844)	**Pav.**
Pavón y Jiménez, José Antonio	
vide Pavón, J.A.	**Pav.**
Pax, Ferdinand Albin (1858-1942)	**Pax**
Payot, Vénance Marie (1826-1902)	**Payot**
Pease, Arthur Stanley (1881-1964)	**Pease**
Peck, Charles Horton (1833-1917)	**Peck**
Penas, Ángel (1948-)	**Penas**
Penas Merino, Ángel	
vide Penas, Á.	**Penas**
Penny, George (? -1838)	**Penny**
Pérard, Alexandre Jules César (1834-1887)	**Pérard**
Pereira Coutinho, António Xavier	
vide Coutinho, A.X. Pereira	**Cout.**
Pérez Arbeláez, Enrique (1896-1972)	**Pérez Arbel.**
Pérez Carro, Francisco Javier (1959-)	**Pérez Carro**
Pérez-Latorre, Andrés Vicente (1965-)	**Pérez Lat.**
Pericás, Jóan Josep (1959-)	**Pericás**
Pericás Mestre, Jóan Josep	
vide Pericás, J.J.	**Pericás**
Perkins, Janet Russel (1853-1933)	**J.R.Perkins**
Perleb, Karl (Carl) Julius (1794-1845)	**Perleb**
Perrottet, George (Georges Guerrard) Samuel (1793-1870)	**Perr.**
Perry, Henry S. (fl. 1860s)	**H.S.Perry**
Perry, Lily May (1895-1992)	**L.M.Perry**
Peter, (Gustav) Albert (1853-1937)	**Peter**
Petit, François Pourfour du (1664-1741)	**P.F.Petit**
Petitmengin, Marcel Georges Charles (1881-1908)	**Petitm.**
Petit-Thouars, Louis-Marie Aubert Aubert Du	
vide Du Petit-Thouars, L.-M.A. Aubert	**Thouars**
Petiver, James (1658-1718)	**Petiver**

Petrov, Vsevolod Alexeevič (1896-1955)	**Petrov**
Pfeiffer, Norma Etta (1889- ?)	**N.Pfeiff.**
Pfitzer, Ernst Hugo Heinrich (1846-1906)	**Pfitzer**
Philippi, Rudolph (Rudolf) Amandus (later Rodolfo Amando) (1808-1904)	**Phil.**
Philipson, William Raymond (1911-)	**Philipson**
Phillips, William Henry (1830-1923)	**W.H.Phillips**
Pia, Julius von (1887-1943)	**Pia**
Pichi-Sermolli, Rodolfo Emilio Giuseppe	
vide Pichi Sermolli, R.E.G.	**Pic.Serm.**
Pichi Sermolli, Rodolfo Emilio Giuseppe (1912-)	**Pic.Serm.**
Picot de la Peyrouse, Philippe	
vide Lapeyrouse, P. Picot de	**Lapeyr.**
Pignatti, Sandro (Alessandro) (1930-)	**Pignatti**
Pinto, Manuel Cabral de Resende-	
vide Resende-Pinto, M.C. de	**Res.-Pinto**
Pinto da Silva, António Rodrigo	
vide Silva, A.R. Pinto da	**P.Silva**
Pirotta, Pietro Romualdo (1853-1936)	**Pirotta**
Piso, Willem (1611-1678)	**Piso**
Pitot, Albert Auguste Louis (1905-)	**Pitot**
Pittier de Fábrega, Henri (Henry) François	
vide Pittier, H.F.	**Pittier**
Pittier, Henri (Henry) François (1857-1950)	**Pittier**
Pitton de Tournefort, Joseph	
vide Tournefort, J. Pitton de	**Tourn.**
Planchon, Jules Émile (1823-1888)	**Planch.**
Plukenet, Leonard [latine Plukenetius] (1642-1706)	**Pluk.**
Plumier, Charles (1646-1704)	**Plum.**
Pocock, Mary Agard (1886-1977)	**Pocock**
Podpěra, Josef (1878-1954)	**Podp.**
Poeppig, Edouard Friedrich (1798-1868)	**Poepp.**
Poggenburg, Justus Ferdinand (1840-1893)	**Poggenb.**
Pohl, Johann (Baptist) Emanuel (1782-1834)	**Pohl**
Poirault, Marie Henri Georges (1858-1936)	**G.Poirault**
Poiret, Jean Louis Marie (1755-1834)	**Poir.**
Pojarkova, Antonina Ivanovna (1897-1980)	**Pojark.**
Pokorny, Alois (Aloys) (1826-1886)	**Pokorny**
Pollard, Charles Louis (1872-1945)	**Pollard**
Pollini, Ciro (Cyrus) (1782-1833)	**Pollini**
Ponce de León y Aimé, Antonio (1887-1961)	**A.Ponce de León**
Ponce, Marta Monica (1954-)	**Ponce**
Pong, Sin-Min (fl. 1932)	**Pong**
Porsild, Alf Erling (1901-1977)	**A.E.Porsild**
Porsild, Morton Pedersen (1872-1956)	**Porsild**
Portenschlag-Ledermayer, Franz von (1772-1822)	**Port.-Led.**

Porter, Thomas Conrad (1822-1901)	**Porter**
Pospichal, Eduard (1838-1905)	**Posp.**
Posthumus, Oene (1898-1945)	**Posth.**
Potts, Thomas Henry (1824-1888)	**Potts**
Pouzar, Zdeněk (1932-)	**Pouzar**
Poyser, William Aldworth (1882-1928)	**Poyser**
Prada, Maria del Carmen Isabel (1953-)	**Prada**
Prada Moral, Maria del Carmen Isabel	
vide Prada, M.d.C.I.	**Prada**
Prado, Jefferson (1964-)	**Prado**
Praeger, Robert Lloyd (1865-1953)	**Praeger**
Prager, Ernst (Ernest) (1866-1913)	**Prag.**
Prain, David (1857-1944)	**Prain**
Prantl, Karl Anton Eugen (1849-1893)	**Prantl**
Pray, Thomas Richard (1923-)	**Pray**
Preissmann, Ernst (1844- ?)	**Preissm.**
Prelli, Rémy (1947-)	**Prelli**
Prentice, Charles Brightly (1820-1894)	**Prent.**
Presl, Carl (Karl, Carel, Carolus) Bořivoj (Bořiwog,	
Bořiwag) (1794-1852)	**C.Presl**
Presl, Jan Svatopluk (Swatopluk) (1791-1849)	**J.Presl**
Price, Michael Greene (1941-)	**M.G.Price**
Prieto, José Antonio Fernández	
vide Fernández Prieto, J.A.	**Fern.Prieto**
Prince, Arthur Reginald (1900-1969)	**A.Prince**
Pritchard, Stephen F. (fl. 1836)	**S.Pritch.**
Pritzel, Ernst Georg (1875-1946)	**E.Pritz.**
Proctor, George Richardson (1920-)	**Proctor**
Prouvençal (Prouvensal, Provençal) de Saint-Hilaire,	
Auguste (Augustin) François César	
vide Saint-Hilaire, A.F.C. Prouvençal de	**A.St.-Hil.**
Pryer, Kathleen Mary (1955-)	**Pryer**
Pulle, August Adriaan (1878-1955)	**Pulle**
Punetha, Nilamber (1949-)	**Punetha**
Pursh, Friedrick (Friedrich) Traugott (né Pursch, F.T.) (1774-1820)	**Pursh**
Pylaie, Auguste Jean Marie Bachelot de la	
vide Bachelot de la Pylaie, A.J.M.	**Bach.Pyl.**
Pynaert, Charles (1872-1936)	**Pynaert**
Qian, Yi-Yong (1930-)	**Y.Y.Qian**
Qin, Ren-Chang	
vide Ching, Ren-Chang	**Ching**
Qiu, Pei-Xi	
vide Chiu, Pei-Shi	**P.S.Chiu**
Quansah, Nathaniel (1953-)	**Quansah**
Quézel, Pierre Ambrunaz (1926-)	**Quézel**

Quirk, Helen Margaret (1953-1982)	**H.M.Quirk**
Rabenhorst, Gottlob (Gottlieb) Ludwig (1806-1881)	**Rabenh.**
Raciborski (Raciborsky), Marjan (Marian, Maryan, Maryjam) (1863-1917)	**Racib.**
Raddi, Giuseppe (1770-1829)	**Raddi**
Rae, Frederick James (1883-1941)	**Rae**
Raffenau-Delile, Alire	
vide Delile, A. Raffenau	**Delile**
Rafinesque, Constantin Samuel (1783-1840)	**Raf.**
Rafinesque-Schmaltz, Constantin Samuel	
vide Rafinesque, C.S.	**Raf.**
Raimann, Rudolf (1863-1896)	**Raim.**
Raine, Catherine Ann (1968-)	**Raine**
Raizada, Mukat Behari (1907-)	**Raizada**
Rakotondrainibe, France Nicole Marie (née Michel, F.N.M.) (1944-)	**Rakotondr.**
Ranker, Thomas Anthony (1952-)	**Ranker**
Rankin, Josephine Margaret (née Camus, J.M.) (1949-)	**J.M.Rankin**
Rao, L.N. (fl. 1944)	**L.N.Rao**
Rao, Ramachandra Raghavendra (1945-)	**R.R.Rao**
Raoul, Édouard Fiacre Louis (1815-1852)	**Raoul**
Rapp, William F. Jr. (1918-)	**W.F.Rapp**
Rasbach, Helga (1924-)	**Rasbach**
Rasbach, Kurt (1923-)	**K.Rasbach**
Rauh, Werner (1913-)	**Rauh**
Rauschert, Stephan (1931-1986)	**Rauschert**
Ravi, N. (fl. 1969-1979)	**Ravi**
Rawson, Rawson William (1812-1899)	**Rawson**
Ray, John (until 1669 Wray) [latine Raius, Rajus] (1627-1705)	**Ray**
Raymond, Marcel (Louis-Florent-Marcel) (1915-1972)	**Raymond**
Rebassa, Antoni (1968-)	**Rebassa**
Rebassa Ordinas, Antoni	
vide Rebassa, A.	**Rebassa**
Rechinger, Karl Heinz (1906-)	**Rech.f.**
Reed, Clyde Franklin (1918-)	**C.F.Reed**
Rees, Abraham (1743-1825)	**Rees**
Reeves, Timothy (1947-)	**T.Reeves**
Regan, Margaret (1941-)	**Regan**
Regel, Eduard (August) von (1815-1892)	**Regel**
Reichard, Johann Jakob (Jacob) (1743-1782)	**Reichard**
Reichardt, Heinrich Wilhelm (1835-1885)	**Reichardt**
Reichenbach, (Heinrich Gottlieb) Ludwig (pater) (1793-1879)	**Rchb.**
Reichenbach, Heinrich Gustav (filius) (1824-1889)	**Rchb.f.**
Reichstein, Tadeus (1897-)	**Reichst.**
Reimers, Hermann (Johann Otto) (1893-1961)	**Reimers**

Reinecke, Franz (1866- ?)	**Reinecke**
Reinsch, Paul Friedrich (1836-1914)	**Reinsch**
Reinwardt, Caspar (Kaspar) Georg Carl (Karl) (1773-1854)	**Reinw.**
Rémy, Ezechiel Jules (1826-1893)	**J.Rémy**
Rendle, Alfred Barton (1865-1938)	**Rendle**
Rensch, Ilse (1902-)	**Rensch**
Rensch-Maier, Ilse	
vide Rensch, I.	**Rensch**
Requien, Esprit (1788-1851)	**Req.**
Resende-Pinto, Manuel Cabral de (1912-1989)	**Res.-Pinto**
Retzius, Anders Jahan (1742-1821)	**Retz.**
Reuter, George (Georges) François (1805-1872)	**Reut.**
Reveal, James Lauritz (1941-)	**Reveal**
Rey-Pailhade, Constantin de (1844-1930)	**Rey-Pailh.**
Rezende-Pinto, Manuel Cabral de	
vide Resende-Pinto, M.C. de	**Res.-Pinto**
Reznicek, Anton Albert (1950-)	**Reznicek**
Rheede tot Draakestein, Hendrik Adrian van (1637-1691)	**Rheede**
Riba, Ramón (1934-)	**Riba**
Richard, Achille (1794-1852)	**A.Rich.**
Richard, Louis Claude Marie (1754-1821)	**Rich.**
Richardson, John (1787-1865)	**Richardson**
Richter, Vincenz Aladár (1868-1927)	**V.A.Richt.**
Rico, Enrique (1953-)	**E.Rico**
Rico Hernández, Enrique	
vide Rico, E.	**E.Rico**
Ridley, Henry Nicholas (1855-1956)	**Ridl.**
Ridlon, Harry Cooper (1885-1942)	**Ridlon**
Riley, John (c.1796-1846)	**Riley**
Ritgen, Ferdinand August Maria Franz von (1787-1867)	**Ritgen**
Ritter-Studnička, Hilda (1911-1976)	**Ritter-Studn.**
Rivas Martínez, Salvador (1935-)	**Rivas Mart.**
Rivas-Martínez, Salvador	
vide Rivas Martínez, S.	**Rivas Mart.**
Robinson, Benjamin Lincoln (1864-1935)	**B.L.Rob.**
Robinson, John (1846-1925)	**J.Rob.**
Robinson, Winifred Josephine (1867- ?)	**W.J.Rob.**
Robyns, Frans Hubert Edouard Arthur Walter (1901-1986)	**Robyns**
Rodin, Robert Joseph (1922-1978)	**R.J.Rodin**
Rödl-Linder, Gisela (1954-)	**Rödl-Lind.**
Rodríguez Rios, Roberto (1944-)	**R.Rodr.**
Rodway, Leonard (1853-1936)	**Rodway**
Roehling, Johann Christoph	
vide Röhling, J.C.	**Röhl.**
Roemer, Carl (Karl) Ferdinand von (1818-1891)	**F.Roem.**
Roemer, (Friedrich) Adolph (1809-1869)	**A.Roem.**

Roemer, Johann Jakob (Jacob) (1763-1819)	**Roem.**
Roemer, Max Joseph (1791-1849)	**M.Roem.**
Roeper, Johannes August Christian (1801-1885)	**Roep.**
Rohlena, Josef (Joseph) (1874-1944)	**Rohlena**
Röhling, Johann Christoph (1757-1813)	**Röhl.**
Roivainen, Heikki (1900-1983)	**Roiv.**
Rojas Acosta, Nicolás (1873-1947)	**Rojas Acosta**
Rojas Clemente, Simón de	
vide Clemente y Rubio, S. de Rojas (Roxas)	**Clemente**
Rolleri, Cristina (Hilda) (1945-)	**Rolleri**
Rolleri de Dougherty, Cristina (Hilda)	
vide Rolleri, C.(H.)	**Rolleri**
Romariz, Carlos Mateus (1920-)	**Romariz**
Ronniger, Carl (Karl) (1871-1954)	**Ronniger**
Roos, Marco Cornelis (1955-)	**M.C.Roos**
Rose, Joseph Nelson (1862-1928)	**Rose**
Rosendahl, Henrik Viktor (1855-1918)	**H.Rosend.**
Rosenstock, Eduard (1856-1938)	**Rosenst.**
Ross, Hermann (1862-1942)	**H.Ross**
Rosselló, Josep Antoni (1961-)	**Rosselló**
Rossi, Ludwig (Ljudevit) (1850-1932)	**L.Rossi**
Roth, Albrecht Wilhelm (1757-1834)	**Roth**
Rothmaler, Werner Hugo Paul (1908-1962)	**Rothm.**
Rothrock, Joseph Trimble (1839-1922)	**Rothr.**
Rousseau, (Joseph Jules Jean) Jacques (1905-1970)	**J.Rousseau**
Roux, Jacobus Petrus (1954-)	**J.P.Roux**
Rouy, Georges (C.C.) (1851-1924)	**Rouy**
Rovirosa, José N. (1849-1901)	**Rovirosa**
Roxas Clemente y Rubio, Simón de	
vide Clemente y Rubio, S. de Rojas (Roxas)	**Clemente**
Roxburgh, William (1751-1815)	**Roxb.**
Roy, R.P. (fl. 1962)	**R.P.Roy**
Roy, Sisir Kumar (1928-)	**S.K.Roy**
Royen, Adriaan van (1704-1779)	**Royen**
Rudge, Edward (1763-1846)	**Rudge**
Rudmose Brown, Robert Neal	
vide Brown, R.N.R.	**R.N.R.Br.**
Rudolphi, (Israel) Karl Asmund [Carl Asmunt (Asmus)]	
(1771-1832)	**Rudolphi**
Rugg, Harold Goddard (1883-1957)	**Rugg**
Ruiz López, Hipólito (1754-1815)	**Ruiz**
Rumpf, Georg Eberhard (Everhard) [latine Rumphius]	
(1627/28-1702)	**Rumph.**
Runemark, Hans (1927-)	**Runemark**
Rupin, Ernest (Jean Baptiste) (1845-1909)	**Rupin**
Ruppius (Rupp), Heinrich Bernhard (1688-1719)	**Ruppius**

Ruprecht, Franz Josef (Ivanovich) (1814-1870)	**Rupr.**
Rury, Phillip MacDonald (1952-)	**Rury**
Rush, Richard John (1944-)	**Rush**
Russow, Edmund (August Friedrich) (1841-1897)	**Russow**
Rydberg, Pehr (Per) Axel (1860-1931)	**Rydb.**
Rylands, Thomas Glazebrook (1818-1900)	**Rylands**
Sabransky, Heinrich (1864-1915)	**Sabr.**
Saccardo, Pier Andrea (1845-1920)	**Sacc.**
Sachet, Marie-Hélène (1922-1986)	**Sachet**
Sadebeck, Richard Emil Benjamin (1839-1905)	**Sadeb.**
Sadler, Joseph (1791-1849)	**Sadler**
Sáenz de Rivas, Concepción (née Sáenz Laín, C.) (1935-)	**Sáenz de Rivas**
Sáenz Laín, Concepción (later Sáenz de Rivas, C.) (1935-)	**Sáenz Laín**
Sagawa, Noboru (1920-)	**Sagawa**
Sagorski, Ernest Adolf (1847-1929)	**Sagorski**
Sahashi, Norio (1940-)	**Sahashi**
Saiki, Yasuhisa (1928-)	**Saiki**
Saint-Hilaire, Auguste (Augustin) François César Prouvençal de (1779-1853)	**A.St.-Hil.**
Saint-Lager, Jean Baptiste (1825-1912)	**St.-Lag.**
Saint-Pierre, Jacques Nicolas Ernest Germain de *vide* Germain de Saint-Pierre, J.N.E.	**Germ.**
Sakurai, Kyuichi (1889-1963)	**Sakurai**
Saldanha, Cecil John (1926-)	**C.J.Saldanha**
Salgado, Arthur Edward (1946-)	**Salgado**
Salis-Marschlins, Carl Ulysses Adalbert von (1795-1886)	**Salis**
Salisbury, Richard Anthony (né Markham, R.A.) (1761-1829)	**Salisb.**
Salleh, Kamarudin Mat (1959-)	**Salleh**
Salomon, Carl (E.) (1829-1899)	**Salomon**
Salt, Henry (1780-1827) *	**Salt**
Salvo, Ángel Enrique (1957-)	**Salvo**
Salvo Tierra, Ángel Enrique *vide* Salvo, Á.E.	**Salvo**
Samant, Sher Singh (1962-)	**Samant**
Sampaio, Alberto José de (1881-1946)	**A.Samp.**
Sampaio, Gonçalo António da Silva Ferreira (1865-1937)	**Samp.**
Sánchez Villaverde, Carlos Alfredo (1953-)	**C.Sánchez**
Sander, Henry Frederick Conrad (1847-1920)	**Sander**
Sandford, E. (fl. 1882)	**Sandford**
Sandmark, Gudmund (fl. 1809)	**Sandmark**
Sándor, J. (fl. 1860s) *	**Sándor**
Sanio, Carl (Karl) Gustav (1832-1891)	**Sanio**
Saraiva Batarda, Rosette Mercedes *vide* Batarda, Rosette Mercedes Saraiva	**Batarda**
Sarnthein, Ludwig von (1861-1914)	**Sarnth.**

49

Sarnthein zu Rottenburg, Kellerburg und Kränzelstein, Ludwig von	
vide Sarnthein, L. von	**Sarnth.**
Sarvela, Jaakko (1914-)	**Sarvela**
Sasaki, Syun'iti (Shun-ichi, Syuniti) (1888-1960)	**Sasaki**
Sastre, Bartomeu (1956-)	**B.Sastre**
Sastre, Claude Henri Léon (1938-)	**Sastre**
Sastre Palou, Bartomeu	
vide Sastre, B.	**B.Sastre**
Sastry, Addala RamaKrishna (1938-)	**Sastry**
Satake, Yoshisuke (1902-)	**Satake**
Satija, Chander Kanta (née Trikha, C.K.) (1942-)	**Satija**
Satô, Kunihiko (Kunio) (1902-)	**K.Satô**
Satomi, Nobuo (1922-)	**Satomi**
Satou, Zyun (1957-)	**Satou**
Sauget-Barbier, Joseph Sylvestre	
vide Léon, Frère	**Léon**
Sauget y Barbis, Joseph Sylvestre	
vide Léon, Frère	**Léon**
Saunders, Richard Mark Kingsley (1964-)	**R.M.K.Saunders**
Sauter, Anton Eleutherius (1800-1881)	**Saut.**
Sauvages de la Croix, François Boissier de	
vide Sauvages, F. Boissier de la Croix de	**Sauvages**
Sauvages, François Boissier de la Croix de (1706-1767)	**Sauvages**
Sauzé, Jean Charles (1815-1889)	**Sauzé**
Savatier, Paul Amadée Ludovic (1830-1891)	**Sav.**
Savigny, (Marie) Jules César Lélorgne de (1777-1851)	**Savigny**
Săvulescu, Traian (Trajan) (1889-1963)	**Săvul.**
Scamman, Edith (Henry) (1882-1967)	**Scamman**
Schaffner, John Henry (1866-1939)	**J.H.Schaffn.**
Schaffner, Wilhelm (later Johann Guillermo) (1830-1882)	**W.Schaffn.**
Schauer, Sebastian (fl. 1847)	**S.Schauer**
Schelpe, Edmund André Charles Louis Eloi (1924-1985)	**Schelpe**
Schenck, (Johann) Heinrich (Rudolf) (1860-1927)	**Schenck**
Schidlay, Eugen (1911-)	**Schidlay**
Schinz, Hans (1858-1941)	**Schinz**
Schippers, Rijke Roelf (1943-)	**Schippers**
Schkuhr, Christian (1741-1811)	**Schkuhr**
Schlechtendal, Diederich Franz Leonhard von (1794-1866)	**Schltdl.**
Schleicher, Johann Christoph (1768-1834)	**Schleich.**
Schlosser von Klekovski, Joseph Calasenz (Jossib Calasancij)	
(1808-1882)	**Schloss.**
Schmid, Rupertus (? -1804)	**Rup.Schmid**
Schmidel, Casimir Christoph (1718-1792)	**Schmidel**
Schmidt, Justus J.H. (1851-1930)	**J.J.H.Schmidt**
Schmidt, Otto Christian (1900-1951)	**O.C.Schmidt**

Schmiedel, Casimir Christoph
   vide Schmidel, C.C.   **Schmidel**
Schmitz, Johann Joseph (1813-1845)   **J.J.Schmitz**
Schneider, George (1848-1917)   **G.Schneid.**
Schneider, Ulrike (1936-   )   **U.Schneid.**
Schneller, Johann Jakob (1942-   )   **Schneller**
Schoenefeld, Wladimir de (1816-1875)   **Schoenef.**
Scholtz, Johann Eduard Heinrich (1812-1859)   **H.Scholtz**
Schomburgk, Robert Hermann (1804-1865)   **R.H.Schomb.**
Schott, Heinrich Wilhelm (1794-1865)   **Schott**
Schrader, Heinrich Adolph (1767-1836)   **Schrad.**
Schrank, Franz von Paula von (1747-1835)   **Schrank**
Schreber, Johann Christian Daniel von (1739-1810)   **Schreb.**
Schube, Theodor (1860-1934)   **Schube**
Schultes, Joseph August (1773-1831)   **Schult.**
Schultz, Carl (Karl) Friedrich (1765/66-1837)   **Schultz**
Schultz, Carl (Karl) Heinrich
   vide Schultz-Schultzenstein, C.H.   **Schultz-Sch.**
Schultz, Friedrich Wilhelm (1804-1876)   **F.W.Schultz**
Schultz-Schultzenstein, Carl (Karl) Heinrich (1798-1871)   **Schultz-Sch.**
Schumacher, Albert (1893-1975)   **A.Schumach.**
Schumacher, (Henrich) Christian Friedrich (1757-1830)   **Schumach.**
Schumann, Eva (1889- ?)   **E.Schum.**
Schumann, Karl Moritz (1851-1904)   **K.Schum.**
Schur, Philipp Johann Ferdinand (1799-1878)   **Schur**
Schwacke, Carl August Wilhelm (1848-1904)   **Schwacke**
Schwarz, Otto Karl Anton (1900-1983)   **O.Schwarz**
Schweinfurth, Georg August (1836-1925)   **Schweinf.**
Scoggan, Homer John (1911-1986)   **Scoggan**
Scopoli, Giovanni Antonio (Johannes Antonius) (1723-1788)   **Scop.**
Scott, John (1838-1880)   **J.Scott**
Scott, Robert Robinson (1827-1877)   **R.R.Scott**
Scriba, Julius Karl (1848-1905)   **J.Scriba**
Seaton, Henry Eliason (1869-1893)   **Seaton**
Seelos, Gustav (1831-1911)   **Seelos**
Seemann, Berthold Carl (1825-1871)   **Seem.**
Séguier, Jean François (1703-1784)   **Ség.**
Sehnem, Aloysio (1912-1981)   **Sehnem**
Seiler, Ralph Leslie (1948-   )   **R.L.Seiler**
Seitz, Wolfgang (1940-   )   **W.Seitz**
Selander, Nils Sten Edward (1891-1957)   **Selander**
Selett, J.W. (fl. 1941)   **Selett**
Selling, Olof Hugo (1917-   )   **Selling**
Sen, Tuhinsri (Tuhinseri) (1942-   )   **T.Sen**
Sen, Udayananda (1933-   )   **U.Sen**
Senna, Rosana Moreno (1967-   )   **R.M.Senna**

Sennen, Frère (né Granier-Blanc, Étienne Marcellin) (1861-1937)	**Sennen**
Serizawa, Shunsuke (1948-)	**Seriz.**
Sessé, Martin de	
vide Sessé y Lacasta, M.	**Sessé**
Sessé y Lacasta, Martin de (1751-1808)	**Sessé**
Seubert, Moritz August (1818-1878)	**Seub.**
Seward, Albert Charles (1863-1941)	**Seward**
Seymann, Wilhelm (1887-1915)	**Seymann**
Seymour, Frank Conkling (1895-1985)	**F.Seym.**
Shafer, John Adolph (1863-1918)	**Shafer**
Sharma, Anurita (fl. 1992)	**An.Sharma**
Shaver, Jesse Milton (1888-1961)	**Shaver**
Sheffield, Elizabeth (1954-)	**Sheffield**
Shende, D.V. (fl.1945)	**Shende**
Shi, Lei (1967-)	**L.Shi**
Shi, Su-Hua (1956-)	**S.H.Shi**
Shieh, Wang-Chueng (1929-)	**W.C.Shieh**
Shim, Phyau-Soon (1942-)	**P.S.Shim**
Shimek, Bohumil (1861-1937)	**Shimek**
Shimura, Yoshio (1920-)	**Shimura**
Shing, Kung-Hsia (Gung-Hsia, Gong-Hsia, Kung-Shieh) (1929-)	**K.H.Shing**
Shivas, Mary Grant (later Walker, M.G.) (1926-)	**Shivas**
Short, John William (1952-)	**J.W.Short**
Shukla, Pradeep Kumar (1961-)	**P.K.Shukla**
Shuttleworth, Robert James (1810-1874)	**Shuttlew.**
Sidebotham, Joseph (1822-1884)	**Sideboth.**
Sieber, Franz(e) Wilhelm (1789-1844)	**Sieber**
Siegesbech, Johann Georg (1686-1755)	**Siegesb.**
Silliman, Benjamin (1779-1864)	**Silliman**
Silva, António Rodrigo Pinto da (1912-1992)	**P.Silva**
Silveira, Alvaro Astolpho da (1867-1945)	**Silveira**
Sim, Robert (1791-1878)	**R.Sim**
Sim, Thomas Robertson (1856-1938)	**Sim**
Singh, Sarnam (1957-)	**Sarn.Singh**
Singh, Shri Surendra (1938-)	**Sur.Singh**
Sinha, B.M.B. (fl. 1970-1988)	**B.M.B.Sinha**
Siplivinsky (Siplivinskij), Vladimir N. (1937-)	**Sipliv.**
Skog, Judith Ellen (1944-)	**J.E.Skog**
Skottsberg, Carl Johan Fredrik (1880-1963)	**Skottesb.**
Sledge, William Arthur (1904-1991)	**Sledge**
Sleep, Anne (1939-1993)	**Sleep**
Sloane, Hans (1660- 1753)	**Sloane**
Slosson, Margaret (1872- ?)	**Sloss.**
Small, John Kunkel (1869-1938)	**Small**
Smith, Alan Reid (1943-)	**A.R.Sm.**
Smith, Albert Charles (1906-)	**A.C.Sm.**

Smith, Dale Metz (1928-)	**D.M.Sm.**
Smith, Doris Alma (née Goy, D.A.) (1912-)	**D.A.Sm.**
Smith, F. Donnel (? - ?)	**F.Donn.Sm.**
Smith, James (1760-1840)	**Js.Sm.**
Smith, James Edward (1759-1828)	**Sm.**
Smith, Jared Gage (1866-1925)	**J.G.Sm.**
Smith, John (1798-1888)	**J.Sm.**
Smith, John Donnell (1829-1928)	**Donn.Sm.**
Smith, Lyman Bradford (1904-)	**L.B.Sm.**
Smith-Dodsworth, John (fl. 1989) *	**Sm.-Dodsw.**
Sodiro, Luis (Aloysius, Luigi) (1836-1909)	**Sodiro**
Sojàk, Jiří (1936-)	**Sojàk**
Solander, Daniel Carl (1753-1782)	**Sol.**
Solms-Laubach, Hermann Maximilian Carl Ludwig	
Friedrich zu (1842-1915)	**Solms**
Soltis, Douglas E. (fl. 1987)	**D.E.Soltis**
Soltis, Pamela S. (fl. 1987)	**P.S.Soltis**
Somers, Paul (1945-)	**P.Somers**
Sommerfelt, Søren Christian (Severinus Christianus)	
(1794-1838)	**Sommerf.**
Sommier, Carlo Pietro Stefano (Stephen) (1848-1922)	**Sommier**
Sóo, Károli Rezsö	
vide Sóo von Bere, K.R.	**Sóo**
Sóo von Bere, Károli Rezsö (Rudolf) (1903-1980)	**Sóo**
Soper, James Herbert (1916-)	**Soper**
Sota, Elias Ramon de la	
vide de la Sota, E.R.	**de la Sota**
Sowerby, James (1757-1822)	**Sowerby**
Spach, Édouard (1801-1879)	**Spach**
Sparre, Benkt (Bengt) (1918-1986)	**Sparre**
Spenner, Fridolin Carl Leopold (1798-1841)	**Spenn.**
Splitgerber, Frederik Louis (Friedrich Ludwig) (1801-1845)	**Splitg.**
Sprengel, Anton (1803-1851)	**A.Spreng.**
Sprengel, Kurt (Curt, Curtius) (Polycarp Johachim) (1766-1833)	**Spreng.**
Spring, Anton Friedrich (Antoine Frédéric) (1814-1872)	**Spring**
Spruce, Richard (1817-1893)	**Spruce**
Srivastava, Gopal Krishna (1939-)	**G.K.Srivast.**
St. John, Edward Porter (1866-1953)	**E.P.St.John**
St. John, Harold (1892-1991)	**H.St.John**
St. John, Robert Porter (1869-1960)	**R.P.St.John**
Stace, Clive Anthony (1938-)	**Stace**
Standley, Paul Carpenter (1884-1963)	**Standl.**
Stansfield, Abraham (1802-1880)	**Stansf.**
Stansfield, Frederick Wilson (1854-1937)	**F.W.Stansf.**
Stearn, William Thomas (1911-)	**Stearn**
Steele, Edward Strieby (1850-1942)	**E.S.Steele**

Steele, William Edward (1816-1883)	**Steele**
Steenis, Cornelis Gijsbert Gerrit Jan van (1901-1986)	**Steenis**
Stepanov, Nikolay Vitalievich (1966 -)	**Stepanov**
Sternberg, Caspar (Kaspar) Maria von (1761-1838)	**Sternb.**
Sterns, Emerson Ellick (1846-1926)	**Sterns**
Steudel, Ernst Gottlieb von (1783-1856)	**Steud.**
Stewart, Ralph Randles (1890-1993)	**R.R.Stewart**
Steyermark, Julian Alfred (1909-1988)	**Steyerm.**
Stokes, Jonathan S. (1755-1831)	**Stokes**
Stolze, Robert Gardner (1927-)	**Stolze**
Stone, Benjamin Clemens Masterman (1933-1994)	**B.C.Stone**
Stowell, Willard Allen (? -1929)	**Stowell**
Strasburger, Eduard Adolf (1844-1912)	**Strasb.**
Straszewski, Heinrich von (1887-1944)	**Strasz.**
Strauss, Susanna Elizabeth (née van Schalkwyk, S.E.) (1941-)	**S.E.Strauss**
Strempel, Johannes Karl (Carl) Friedrich (1800-1872)	**Strempel**
Strobl, Gabriel (1846-1925)	**Strobl**
Sturm, Jacob (Jakob) (1771-1848)	**Sturm**
Sturm, Johann Wilhelm (1808-1865)	**J.W.Sturm**
Sudre, Henri (L.) (1862-1918)	**Sudre**
Suessenguth, Karl (1893-1955)	**Suess.**
Sugimoto, Junichi (1901-)	**Sugim.**
Suksdorf, Wilhelm Nikolaus (1850-1932)	**Suksd.**
Summerfelt, Søren Christian	
vide Sommerfelt, S. Ch.	**Sommerf.**
Summers, Lucia A. (1839-1898)	**Summers**
Sumnevicz (Sumnevitcz), Georgij Prokopievic (1909-1947)	**Sumnev.**
Sunding, Per (1937-)	**Sunding**
Susaki, Chusuke (1866-1933)	**Susaki**
Suzuki, Hyoji (1915-)	**H.Suzuki**
Suzuki, Kiyoshi (fl. 1982)	**K.Suzuki**
Suzuki, Tokio (1911-)	**T.Suzuki**
Svenson, Henry Knute (Knut) (1897-1986)	**Svenson**
Sventenius, Eric R. Svensson (1910-1973) ✳	**Svent.**
Swartz, Olof (Peter) (1760-1818)	**Sw.**
Sweet, Robert (1783-1835)	**Sweet**
Syme, John Thomas Irvine (Irwine) (later Boswell-Syme, J.T.I.)	
(1822-1888)	**Syme**
Symons, Jelinger (1778-1851)	**Symons**
Szyszylowicz, Ignaz (Ignacy) von (1857-1910)	**Szyszyl.**
Tagawa, Motozoi (1908-1977)	**Tagawa**
Tait, A. (fl. 1860s)	**Tait**
Takamine, Noboru (1888-1970)	**Takamine**
Takeda, Hisayoshi (1883-1972)	**Takeda**
Takeuchi, Masayuki (1925-)	**M.Takeuchi**

Takhtajan, Armen Leonovich (1910-)	**Takht.**
Takiguchi, Keiko (later Muro, K.) (1954-)	**Takig.**
Tan, Benito Ching (1948-)	**B.C.Tan**
Tanger, Louise Forney Arnold (1889- ?)	**Tanger**
Tardieu, Marie-Laure	
vide Tardieu-Blot, M.-L.	**Tardieu**
Tardieu-Blot, Marie-Laure (1902-)	**Tardieu**
Taschner, Christian Friedrich (1817- ?)	**Taschner**
Tate, Donald Eugene (1933-)	**D.E.Tate**
Tateishi, Yoichi (1948-)	**Tateishi**
Tatewaki, Misao (1899-)	**Tatew.**
Taton, Auguste (1914-1989)	**Taton**
Taubert, Paul Hermann Wilhelm (1862-1897)	**Taub.**
Tausch, Ignaz Friedrich (1793-1848)	**Tausch**
Tavel, (Rudolf) Franz von (1863-1941)	**Tavel**
Taylor, Roy Lewis (1932-)	**Roy Taylor**
Taylor, Thomas Mayne Cunninghame (1904-1983)	**T.M.C.Taylor**
Taylor, William Carl (1946-)	**W.C.Taylor**
Teijsmann, Johannes Elias (1809-1882)	**Teijsm.**
Tenison-Woods, Julian Edmund (1832-1889)	**Ten.-Woods**
Tenore, Michele (1780-1861)	**Ten.**
Terracciano, Achille (1861-1917)	**A.Terrac.**
Terracciano, Nicola (1837-1921)	**N.Terrac.**
Terry, Emily (née Hitchcok, E.) (1837-1921)	**Terry**
Tetrick, Robert Marshall [II] (1929-1950)	**Tetrick**
Teysmann, Johannes Elias	
vide Teijsmann, J.E.	**Teijsm.**
Thellung, Albert (1881-1928)	**Thell.**
Thieret, John William (1926-)	**Thieret**
Thiselton-Dyer (Thistleton-Dyer), William Turner (1843-1928)	**Dyer**
Thompson, Rufus Henney (1908-1980)	**R.H.Thomps.**
Thomson, George Malcolm (1849-1933)	**G.M.Thomson**
Thore, Jean (1762-1823)	**Thore**
Thouars, Louis-Marie Aubert Aubert Du Petit-	
vide Du Petit-Thouars, L.-M.A. Aubert	**Thouars**
Thouin, André (1747-1824)	**Thouin**
Thunberg, Carl Peter (Pehr) (1743-1828)	**Thunb.**
Thurn, Everard Ferdinand im (1852-1932)	**Thurn**
Thwaites, George Henry Kendrick (1812-1882)	**Thwaites**
Tidestrom, Ivar (Frederick) (né Tidestrøm) (1864-1956)	**Tidestr.**
Tindale, Mary Douglas (1920-)	**Tindale**
Tineo, Vincenzo (1791-1856)	**Tineo**
Titford, William Jowett (Jowit) (1784-1823/7)	**Titford**
Todaro, Agostino (1818-1892)	**Tod.**
Tolentino, Danielo B. (fl. 1987)	**Tolentino**

Tolmachew (Tolmachev, Tolmacev), Alexandr Innokentevich (Innokent'evich) (1903-1979)	**Tolm.**
Tongiorgi, Ezio (1913-1987)	**Tongiorgi**
Toni, Giovanni Battista De	
vide De Toni, G.B.	**De Toni**
Tonini, Carlo (1803-1877)	**Tonini**
Torres (Torres), Néstor (1949-)	**N.Torres**
Torrey, John (1796-1873)	**Torr.**
Tournefort, Joseph Pitton de (1656-1708)	**Tourn.**
Tourte, Yves (1937-)	**Tourte**
Toyokuni, Hideo (1932-)	**Toyok.**
Trabut, Louis (Charles) (1853-1929)	**Trab.**
Trautvetter, Ernst Rudolf von (1809-1889)	**Trautv.**
Trelease, William (1857-1945)	**Trel.**
Treub, Melchior (1851-1910)	**Treub**
Treviranus, Ludolph (Ludolf) Christian (1779-1864)	**Trevir.**
Trevisan de Saint-Léon, Vittore (Victor, Benedetto, Antonio) (1818-1897)	**Trevis.**
Triana, José Jerónimo (1828-1890)	**Triana**
Triana Silva, José Jerónimo	
vide Triana, J.J.	**Triana**
Trikha, Chander Kanta (later Satija, C.K.) (1942-)	**Trikha**
Trimen, Henry (1843-1896)	**Trimen**
Tripathi, Anil Kumar (1959-)	**A.K.Tripathi**
Trivedi, Bhim Shankar (1923-)	**Trivedi**
Troll, (Julius Georg Hubertus) Wilhelm (1897-1978)	**Troll**
Trotter, E.W. (fl. 1880-1890)	**E.W.Tryon**
Trudell, Harry William (1879-1964)	**Trudell**
Tryon, Alice Faber (née Faber, A.) (1920-)	**A.F.Trotter**
Tryon, Rolla Milton (1916-)	**R.M.Tryon**
Tsai, Jinn-Lai (Jenn-Lai, Jen-Lai) (fl. 1977-1994)	**J.L.Tsai**
Tsutsumi, Hisashi (1939-)	**Tsutsumi**
Tsvelev (Tsvelov, Tsvelyov), Nikolai Nikolaievich	
vide Tzvelev, N.N.	**Tzvelev**
Tu, Vu Nguyen (1942-)	**V.N.Tu**
Tuckerman, Edward (1817-1886)	**Tuck.**
Turczaninow, (Porphir Kiril) Nicolai Stepanowitsch von (1796-1864)	**Turcz.**
Turner, Melvin Dennis (1954-)	**M.D.Turner**
Tutcher, William James (1867-1920)	**Tutcher**
Tuyama, Takasi (1910-)	**Tuyama**
Tuzson, János (1870-1943)	**Tuzson**
Tzvelev (Tzvelov, Tsvelev, Tsvelor), Nikolai Nikolaievich (1925-)	**Tzvelev**
Ueno, Tsugimi (1905-1988)	**T.Ueno**
Ugolini, Ugolino (1856-1942)	**Ugolini**
Ulbrich, Oskar Eberhard (1879-1952)	**Ulbr.**

Ule, Ernest Heinrich Georg (1854-1915)	**Ule**
Underwood, Lucien Marcus (1853-1907)	**Underw.**
Unni, K. Sankaran (fl. 1967)	**Unni**
Urban, Ignatz (1848-1931)	**Urb.**
Urville, Jules Sébastien César Dumont d'	
vide Dumont d'Urville, J.S.C.	**d'Urv.**
Usteri, Alfred (1869-1948)	**A.Usteri**
Usteri, Paul (1768-1831)	**Usteri**
Vaccari, Lino (1873-1951)	**Vacc.**
Vahl, Martin (Hendriksen, Henrichsen) (1749-1804)	**Vahl**
Vahl, Martin [II] (1869-1946)	**M.Vahl**
Vaillant, Sébastien (1669-1722)	**Vaill.**
Valdespino, Iván Alberto (1963-)	**Valdespino**
Valdespino Quintero, Iván Alberto	
vide Valdespino, I.A.	**Valdespino**
Valentine, David Henriques (1912-1987)	**Valentine**
van Alderwerelt van Rosenburgh, Cornelis Rugier Willem Karel	
vide Alderwerelt van Rosenburg van, C.R.W.K.	**Alderw.**
van Borssum Waalkes, Jan	
vide Borssum Waalkes, J. van	**Borss.Waalk.**
Van Cat, Do (fl. 1989) ✳	**Van Cat**
van den Berg, Maria Elizabeth	
vide Berg, M.E. van den	**M.E.Berg**
van den Bosch, Roelof Benjamin (1810-1862)	**Bosch**
van den Ende, Willem Pieter	
vide Ende, W.P. van den	**Ende**
van der Werff, Henk (Hendrik, Hessel) (1946-)	**van der Werff**
van Donselaar, Johannes	
vide Donselaar, J. van	**Donsel.**
Van Eseltine, Glen Parker (1888-1938)	**Van Eselt.**
Van Geert, August(e) (1818-1886)	**Van Geert**
van Hall, Herman (Hermanus) Christiaan	
vide Hall, H.C. van	**H.C.Hall**
Van Heurck, Henri Ferdinand (1838-1909)	**Van Heurck**
Van Hoek, Leonardus (1911-)	**Van Hoek**
Van Hoorebeke, Charles Joseph (1790-1821)	**Van Hooreb.**
Van Houtte, Louis Benoît (1810-1876)	**Van Houtte**
van Ooststroom, Simon Jan	
vide Ooststroom, S.J. van	**Ooststr.**
van Steenis, Cornelis Gijsbert Gerrit Jan	
vide Steenis, C.G.G.J. van	**Steenis**
Vareschi, Volkmar (1906-1991)	**Vareschi**
Váróczy, (Váróczky) Edith Cs. (fl. 1963)	**Váróczky**
Vasconcellos, João, de Carvalho e (1897-1972)	**Vasc.**
Vassiljev (Vasilev), Viktor Nikolayevich (1890-1987)	**V.N.Vassil.**

Vasudeva, Surinder Mohan (1944-)	**S.M.Vasudeva**
Vaucher, Jean Pierre Étienne (1763-1841)	**Vaucher**
Veitch, Harry James (1840-1924)	**H.J.Veitch**
Velenovský, Josef (Joseph) (1858-1949)	**Velen.**
Vellozo (Velloso, Veloso), José Mariano da Conceição (1742-1811)	**Vell.**
Vendryès, Albert (? -1916)	**Vendryès**
Ventenat, Étienne Pierre (1757-1808)	**Vent.**
Verdcourt, Bernard (1925-)	**Verdc.**
Verdoorn, Frans (1906-1984)	**Verd.**
Verduyn, Gusbertha Petronella (1960-)	**Verduyn**
Verhey, C.J. (1917-)	**Verhey**
Verma, Sandeep (fl. 1990)	**S.Verma**
Verma, Satish Chander (1931-)	**S.C.Verma**
Verschaffelt, Ambroise Colette Alexandre (1825-1886)	**Verschaff.**
Veselsky, Bedřich (Friedrich) (1813-1866)	**Veselsky**
Viane, Ronald Louis Leo (1951-)	**Viane**
Vicotin, ? (fl. 1945)	**Vicotin**
Victorin, Frère Marie (né Kirouak, Joseph Louis Conrad) (1885-1944)	**Vict.**
Vida, Gábor (1935-)	**Vida**
Vidal, Louis (fl. 1900-1906)	**L.Vidal**
Vidal, Sebastian	
vide Vidal y Soler, S.	**S.Vidal**
Vidal y Soler, Sebastian (1842-1889)	**S.Vidal**
Vieillard, Eugène Deplance Émile (1819-1896)	**Vieill.**
Vignolo-Lutati, Ferdinando (1878-1965)	**Vignolo**
Vigo i Bonada, Josep (1937-)	**Vigo**
Villagrán, Carolina (fl. 1971)	**Villagrán**
Villar, Dominique	
vide Villars, D.	**Vill.**
Villars, Dominique (until 1785 Villar, D.) (1745-1814)	**Vill.**
Visiani, Roberto de	
vide de Visiani, R.	**Vis.**
Viswanathan, M.B. (fl. 1989)	**M.B.Viswan.**
Vitman, Fulgenzio (1728- 1806)	**Vitman**
Vivant, Jean (1923-)	**Vivant**
Viviand-Morel, Joseph Victor (1843-1915)	**Viv.-Morel**
Viviani, Domenico (1772-1840)	**Viv.**
Vogel, Johannes Christian (1963-)	**J.C.Vogel**
Vogler, Johann (Philipp) Andreas (1746-1816)	**J.A.Vogler**
Voigt, Joachim [Johann] Otto (1798-1843)	**Voigt**
Vol, Charles Edward De	
vide De Vol, C.E.	**De Vol**
Vollmann, Franz (1858-1917)	**Vollm.**
Voroschilov (Voroshilov), Vladimir N. (1908-)	**Vorosch.**
Vorster, Pieter Johannes (1945-)	**Vorster**

Vriese, Willem Hendrik de
    vide de Vriese, W.H.   **de Vriese**
Vukotinović, Ljudevit (Ludwig) von Farkas (1813 1893)   **Vuk.**
Vul'f, Eugenii Vladimirovich
    vide Wulff, E.V.   **E.Wulff**

Waalkes, Jan van Borssum
    vide Borssum Waalkes, J. van   **Borss.Waalk.**
Wachter, Willem Hendrik (1882-1946)   **Wacht.**
Wade, Arthur Edwin (Edward) (1895-1989)   **A.E.Wade**
Wagner, David Henry (1945-)   **D.H.Wagner**
Wagner, Florence Signaigo (née Signaigo, F.) (1919-)   **F.S.Wagner**
Wagner, Warren Herbert Jr. (1920-)   **W.H.Wagner**
Waha Baillonville, T. de (fl. 1989) *   **Waha**
Wahlenberg, George (after 1804 Göran) (1780-1851)   **Wahlenb.**
Waisbecker, Anton (1835-1916)   **Waisb.**
Wakefield, Norman Arthur (1918-1972)   **N.A.Wakef.**
Walker-Arnott, George Arnott (Arnold)
    vide Arnott, G.A. Walker-   **Arn.**
Walker, Stanley (1924-1985)   **S.Walker**
Walker, Trevor George (1927-)   **T.G.Walker**
Wall, George (1821-1894)   **G.Wall**
Wallich, Nathaniel (Nathanael) (né Wulff, Wolff, Nathan) (1786-1854)   **Wall.**
Wallroth, Carl Friedrich Wilhelm (1792-1857)   **Wallr.**
Walter, Emile (1873-1953)   **E.Walter**
Walter, Kerry Scott (1950-)   **K.S.Walter**
Walter, Thomas (1740-1789)   **Walter**
Wang, Bo-Sun (Besun, Ba-San) (1931-)   **B.S.Wang**
Wang, Chang-Yong (Chung-Yong) (1934-)   **C.Y.Wang**
Wang, Chu-Hao (1923-)   **Chu H.Wang**
Wang, Chung-Hsin (Chung-Hsi) (1923-)   **Chung H.Wang**
Wang, Chung-Ren
    vide Wang, Zhong-Ren   **Z.R.Wang**
Wang, Jian-Zhong (1952-)   **J.Z.Wang**
Wang, Ke-Qing (1968-) *   **K.Q.Wang**
Wang, Pai-Sun (Bai-Sun)
    vide Wang, Bo-Sun   **B.S.Wang**
Wang, Pei-Shan (1936-)   **P.S.Wang**
Wang, Wei (1912-1991)   **W.Wang**
Wang, Xiao-Ying (fl. 1994)   **X.Y.Wang**
Wang, Zhong-Ren (1939-)   **Z.R.Wang**
Wang, Zhu-Hao
    vide Wang, Chu-Hao   **Chu H.Wang**
Wanntorp, Hans-Erik (1940-)   **Wanntorp**
Warburg, Edmund Frederic (1908-1966)   **E.F.Warb.**
Warburg, Otto (1859-1938)   **Warb.**

Warnstorf, Carl (Friedrich E.) (1837-1921)	**Warnst.**
Warshauer, Frederick Richard (1946-)	**Warshauer**
Watanake, Iwao (fl. 1989-1993) ✳	**Watanake**
Watelet, (Jean-François) Adolphe (1811-1879)	**Watelet**
Waters, Campbell Easter (1872-1955)	**Waters**
Waterway, Marcia Jane (1951-)	**Waterway**
Watson, Hewett Cottrell (1804-1881)	**H.C.Watson**
Watson, Sereno (1826-1892)	**S.Watson**
Watt, David Allan Poe (1830-1917)	**Watt**
Watts, William Walter (1856-1920)	**Watts**
Watzel, Kajetán (Cajetán) (1812-1885)	**Watzel**
Wawra, Heinrich (1831-1887)	**Wawra**
Wawra von Fernsee, Heinrich	
vide Wawra, H.	**Wawra**
Weatherby, Charles Alfred (1875-1949)	**Weath.**
Webb, Philip Barker (1793-1854)	**Webb**
Weber, Friedrich (1781-1823)	**F.Weber**
Weber, Georg Heinrich (1752-1828)	**Weber**
Weber, Ulrich (1898-)	**U.Weber**
Webster, Terry R. (1938-)	**T.R.Webster**
Wei, Yun (1937-)	**Y.Wei**
Weigel, Christian Ehrenfried von (1748-1831)	**Weigel**
Weiller, Marc (1880-1945)	**Weiller**
Weiss (Weis, Weisz), Friedrich Wilhelm	
(Fridericus Guilielmus) (1744-1826)	**Weiss**
Welwitsch, Friedrich Martin Josef (1806-1872)	**Welw.**
Wendt, Thomas (Tom) Leighton (1950-)	**T.Wendt**
Wercklé, Karl (Carl) (1860-1924)	**Wercklé**
Werff, Henk van der	
vide van der Werff, H.	**van der Werff**
Werth, Charles Richard (1947-)	**C.R.Werth**
Weselsky, Bedřich (Friedrich)	
vide Veselsky, B.(F.)	**Veselsky**
Wessels Boer, Jan Gerard (1936-)	**Wess.Boer**
Wettstein, Fritz (Friedrich) (1895-1945)	**F.Wettst.**
Wettstein, Richard (1863-1931)	**Wettst.**
Wettstein von Westersheim, Fritz (Friedrich)	
vide Wettstein, F.(F.)	**F.Wettst.**
Wettstein von Westersheim, Richard	
vide Wettstein, R.	**Wettst.**
Wheeler, George Montague (1842-1905)	**G.M.Wheeler**
Wherry, Edgard Theodore (1885-1982)	**Wherry**
White, Cyril Tenison (1890-1950)	**C.T.White**
White, Richard Alan (1935-)	**R.A.White**
Whitmore, S.A. (fl.1991)	**S.A.Whitmore**
Whitwell, George (c.1840-1924)	**G.Whitw.**

Whitwell, William (1839-1920)	**Whitw.**
Wichers, Friderich Hinrich	
vide Wiggers, F.H.	**F.H.Wigg.**
Widén, Carl-Johann (1935-)	**Widén**
Widén, Hely Katriina (fl. 1978)	**H.K.Widén**
Wieffering, Johan Hendrik (1931-1990)	**Wieff.**
Wierzbicki, Piotr (Peter, Petrus) Pawlus [Paulus] (1794-1847)	**Wierzb.**
Wiggers, Fridrich Hindrich (Friedrich Heinrich) (1746-1811)	**F.H.Wigg.**
Wikström, Johan Emanuel (1789-1856)	**Wikstr.**
Wilce, Joan Hubbell (1931-)	**J.H.Wilce**
Wilczek, Ernst (1867-1948)	**Wilczek**
Wildeman, Émile August(e) Joseph	
vide De Wildeman, E.A.J.	**De Wild**.
Wilkes, Charles Henry Davis (1798-1877)	**Wilkes**
Willdenow, Carl Ludwig von (1765-1812)	**Willd.**
Willemet, Pierre Rémi François de Paule (1762-1790)	**P.Willemet**
Williams, Reginald George (1935-)	**R.G.Williams**
Williams, Samuel (1897-1965)	**S.Williams**
Willison, William (c.1806-1875)	**Willison**
Willkomm, Heinrich Moritz (1821-1895)	**Willk.**
Wilms, Friedrick Heinrich (1811-1880)	**F.H.Wilms**
Wilson, Carl Louis (1897-)	**C.L.Wilson**
Wilson, Kenneth Allen (1928-)	**K.A.Wilson**
Wilson, Leonard Richard (1906-)	**L.R.Wilson**
Wimmer, (Christian) Friedrich (Heinrich) (1803-1868)	**Wimm.**
Windham, Michael Dennis (1954-)	**Windham**
Windisch, Paulo Günther (Guenter) (1948-)	**P.G.Windisch**
Winslow, Evelyn James (1870-1949)	**E.J.Winslow**
Wirtgen, Ferdinand Paul (1848-1924)	**F.Wirtg.**
Wirtgen, Philipp Wilhelm (1806-1870)	**Wirtg.**
Withering, William (1741-1799)	**With.**
Woerlein, Georg (1848-1899)	**Woerl.**
Wollaston, Georg Buchanan (1814-1899)	**Woll.**
Wollenweber, Eckhard (1941-)	**E.Wollenw.**
Wong, Kin-Kuang (fl. 1932)	**K.K.Wong**
Wong, Khoon-Meng (1954-)	**K.M.Wong**
Wood, Alphonso (W.) (1810-1881)	**A.W.Wood**
Woods, Julian Edmund Tenison-	
vide Tenison-Woods, J.E.	**Ten.-Woods**
Woolson, Grace A. (1856-1911)	**Woolson**
Woroschilov, Wladimir N.	
vide Voroschilov, V.N.	**Vorosch.**
Woynar, Heinrich Karl (1865-1917)	**Woyn.**
Wright, Charles (Carlos) (1811-1885)	**C.Wright**
Wright, Charles Henry (1864-1941)	**C.H.Wright**

Wu

Wu, Chao-Hung
   vide Wu, Shiew-Hung   **S.H.Wu**
Wu, Cheng-Yih (1916-   )   **C.Y.Wu**
Wu, Shi-Fu (1946-   )   **S.F.Wu**
Wu, Shiew-Hung (1934-   )   **S.H.Wu**
Wu, Shu-Kung (1935-   )   **S.K.Wu**
Wu, Su-Kung (Su-Gong)
   vide Wu, Shu-Kung)   **S.K.Wu**
Wu, Yin-Chan (Yin-Ch'An) (1901-1959)   **Y.C.Wu**
Wu, Yi-Ping (1961-   )   **Y.P.Wu**
Wu, Zhao-Hong
   vide Wu, Shiew-Hung   **S.H.Wu**
Wu, Zheng-Yih
   vide Wu, Cheng-Yih   **C.Y.Wu**
Wulfen, Franz Xavier von (1728-1805)   **Wulfen**
Wulff (Wulf), Eugenii Vladimirovich (Vladimirowich,
   Vladimirovitsch) (1885-1941)   **E.Wulff**
Wünsche, Friedrich Otto (1839-1905)   **Wünsche**

Xia, Qun (1949-   )   **Q.Xia**
Xiang, Li-Lin (1965-   )   **L.L.Xiang**
Xie, Chang-Fu
   vide Hsieh, Chang-Fu   **C.F.Hsieh**
Xie, Yin-Tang
   vide Hsieh, Yin-Tang   **Y.T.Hsieh**
Xing, Gong-Xia
   vide Shing, Kung-Hsia   **K.H.Shing**

Yabe, Yoshitaka (1876-1931)   **Y.Yabe**
Yadav, Ashok Kumar (1953-   )   **A.K.Yadav**
Yamamoto, Akira (fl. 1995)   **A.Yamam.**
Yamamoto, Yoshimatsu (né Wakamori, Y.) (1893-1947)   **Yamam.**
Yang, Chang-You (1928-   )   **Chang Y.Yang**
Yang, Chun-Yu (1948-   )   **C.Y.Yang**
Yao, Guan-Hu (1947-   )   **G.H.Yao**
Yao, K.W.
   vide Yao, Guan-Hu   **G.H.Yao**
Yapp, Richard Henry (1871-1929)   **Yapp**
Yatabe, Ryôkichi (Ruokichi) (1851-1899)   **Yatabe**
Yates, Lorenzo Gordin (1837-1909)   **Yates**
Yatskievych, George Alfred (1957-   )   **Yatsk.**
Yokote, Etsuko (fl. 1977)   **E.Yokote**
Yoshikawa, Naoto (1951-   )   **Yoshik.**
Yü (Yu), Hung-Ts'an (fl. 1986)   **H.T.Yü**
Yu, Shun-Li (1965-   ) *   **S.L.Yu**
Yuncker, Truman George (1891-1964)   **Yunck.**

Zabel, Hermann (1832-1912)   **Zabel**
Zaffran, Jacques (1935-   )   **Zaffran**
Zamora, Prescillano Martinez (1933-   )   **P.M.Zamora**
Zapałowich, Hugo (1852-1917)   **Zapał.**
Zavaro Pérez, Carlos (1967-   )   **Zavaro**
Zeiller, Charles René (1847-1915)   **Zeiller**
Zenker, Jonathan Carl (Karl) (1799-1837)   **Zenker**
Zersi, Elia (1818-1880)   **Zersi**
Zhang, Can-Ming (1963-   )   **C.M.Zhang**
Zhang, Chao-Fang (1923-   )   **C.F.Zhang**
Zhang, Jin-Lun
  vide Chang, Jin-Lun   **J.L.Chang**
Zhang, Lai-Fa (1938-   )   **L.F.Zhang**
Zhang, Li-Bing (1966-   )   **L.B.Zhang**
Zhang, Xian-Chun (1965-   )   **X.C.Zhang**
Zhang, Yu-Long (1932-   ) ✱   **Y.L.Zhang**
Zhong, Ye-Cong (1935-   )   **Y.C.Zhong**
Zhou, Feng-Qin (1951-   )   **F.Q.Zhou**
Zhou, Hou-Gao (1962-   )   **H.G.Zhou**
Zhu, Wei-Ming
  vide Chu, Wei-Ming   **W.M.Chu**
Zimmer, Brigitte (1943-   )   **B.Zimmer**
Zimmermann, W.J. (fl. 1989) ✱   **W.J.Zimm.**
Zimmermann, Walter Max (1892-1980)   **W.Zimm.**
Zink, Michael Johannes (1954-   )   **Zink**
Zinserling, Yurij Dmitrievtch (1894-1938)   **Zinserl.**
Zippelius, Alexander (1797-1828)   **Zipp.**
Zoëga, Johann (1742-1788)   **Zoëga**
Zogg, Emil (1915-   )   **Zogg**
Zollinger, Heinrich (1818-1859)   **Zoll.**
Zumaglini, Antonio Maurizio (1804-1865)   **Zumagl.**

APPENDIX

This Appendix consists of the abbreviations (or surnames in full) of the names of authors of pteridological *novitates* adopted in the past or recent pubblications (*TL-2* included) which differ from the standard forms recommended in *A.P.N.* and in this book.

The abbreviation of author's name is quoted in boldface in the left hand column; the forename(s) followed by the surname are given in Roman type in the central column; and the standard form is cited in boldface in the right hand column.

Abeywick.	Bartholomeusz Aristides Abeywickrama	**Abeyw.**
Achar.	Erik Acharius	**Ach.**
Afz.	Adam Afzelius	**Afzel.**
J.Ag.	Jacob Georg Agardh	**J.Agardh**
Ait.	William Aiton	**Aiton**
A.Albert	Abel Albert	**Albert**
Alboff	Nikolai Michailovich Albov (Alboff)	**Albov**
H.H.Allan	Harry Howard Allan	**Allan**
J.F.R.Almeida	Joseph Francis Raphael d'Almeida	**d'Almeida**
Anders.	Nils Johan Andersson	**Andersson**
N.J.Anders.	Nils Johan Andersson	**Andersson**
Éd.André	Édouard-François André	**André**
Arbeláez	Enrique Pérez Arbeláez	**Pérez Arbel.**
Arnott	George Arnott Walker Arnott	**Arn.**
Arth.	Joseph Charles Arthur	**Arthur**
Aschers.	Paul Ascherson	**Asch.**
Aschers. et Graebn.	Paul Ascherson et Paul Graebner	**Asch. et Graebn.**
Asplund	Erik Asplund	**Aspl.**
Atchins.	[*sphalmate*] William Sackston Atkinson	**Atk.**
Backh.	James Backhouse (filius)	**Backh.f.**
Backhouse	James Backhouse (pater)	**Backh.**
J.Backhouse Jr.	James Backhouse (filius)	**Backh.f.**
Bail.	Frederick Manson Bailey	**F.M.Bailey**
Bailey	Frederick Manson Bailey	**F.M.Bailey**
Bak.	John Gilbert Baker	**Baker**
E.G.Baker	Edmund Gilbert Baker	**Baker f.**
Balbis	Gioanni Battista Balbis	**Balb.**
I.B.Balf.	Isaac Bayley Balfour	**Balf.f.**
Balfour	Isaac Bayley Balfour	**Balf.f.**
Ballard	Francis Ballard	**F.Ballard**

A.J.Bange	Anthelme-Jean Bange	**Bange**
H.H.Bartlett	Harley Harris Bartlett	**Bartlett**
Bast.	Toussaint Bastard	**Bastard**
Battand.	Jules Aimé Battandier	**Batt.**
N. Beadle	Noel Charles William Beadle	**N.C.W.Beadle**
Beauv.	Ambroise Marie François Joseph Palisot de Beauvois	**P.Beauv.**
Becherer	Alfred Becherer	**Bech.**
G.Beck	Günther Beck von Mannagetta und Lërchenau	**Beck**
Bell.	Carlo Antonio Lodovico Bellardi	**Bellardi**
Berg.	Peter Jonas Bergius	**P.J.Bergius**
Bert.	Carlo Luigi Giuseppe Bertero	**Bertero**
Bert.	Antonio Bertoloni	**Bertol.**
L.Bishop	Luther Earl Bishop	**L.E.Bishop**
Blanford	Henry Francis Blanford	**Blanf.**
Blatter	Ethelbert Blatter	**Blatt.**
Bl.	Carl Ludwig von Blume	**Blume**
A.Bobrov	Andrej Eugenievich Bobrov	**A.E.Bobrov**
W.Boer	Jan Gerard Wessels Boer	**Wess.Boer**
J.Boivin	(Joseph) Bernard Boivin	**B.Boivin**
Boj.	Wenceslas Bojer	**Bojer**
Bolt.	James Bolton	**Bolton**
Bomm.	Jean-Édouard Bommer	**J.Bommer**
v.Borss.Waalk.	Jan van Borssum Waalkes	**Borss.Waalk.**
Bowdich	Thomas Edward Bowdich	**T.E.Bowdich**
A.Br.	Alexander Braun	**A.Braun**
Braithw.	Anthony Forrester Braithwaite	**A.F.Braithw.**
Łucy Braun	Emma Lucy Braun	**E.L.Braun**
Britt.	Nathaniel Lord Britton	**Britton**
E.Britt.	Elizabeth Gertrude Britton	**E.Britton**
A.Brongn.	Adolphe Théodore Brongniart	**Brongn.**
Broun	Maurice Broun	**M.Broun**
E.Brown	Elizabeth Dorothy Brown	**E.D.Br.**
F.Brown	Forest Buffen Harkness Brown	**F.Br.**
T.Bryant	Truman Rai Bryant	**T.R.Bryant**
B.S.P.	Nathaniel Lord Britton, Emerson Ellick Sterns et Justus Ferdinand Poggenburg	**Britton, Sterns et Poggenb.**
Bub.	Pietro Bubani	**Bubani**
L.v.Buch	Christian Leopold von Buch	**Buch**
Buckl.	Samuel Botsford Buckley	**Buckley**
N.L.Burm.	Nicolaas Laurens Burman	**Burm.f.**
Busch	Nicolai Adolfowitsch Busch	**N.Busch**
Campbell	Douglas Houghton Campbell	**Campb.**
Carm.	Dugald Carmichael	**Carmich.**

Carr.	Élie-Abel Carrière	Carrière
Carr.	William Carruthers	Carruth.
Carruthers	William Carruthers	Carruth.
Carv.Vasc.	João de Carvalho e Vasconcellos	Vasc.
Celak.	Ladislav Frantisek Celakovský	L.F.Celak.
T.Chambers	Thomas Carrick Chambers	T.C.Chambers
J.I. Chang	[sphalmate] Jin-Lu Chang	J.L.Chang
Chaten.	Constant Chatenier	Chatenier
Chauvin	François Joseph Chauvin	Chauv.
Cheesem.	Thomas Frederic Cheeseman	Cheeseman
Chev.	François Fulgis Chevallier	Chevall.
A.Cheval.	Auguste Jean Baptiste Chevalier	A.Chev.
F.F.Cheval.	François Fulgis Chevallier	Chevall.
Chien	Sung-Shu Chien	S.S.Chien
Chinn.	Robert James Chinnock	Chinnock
Chiu	Pei-Shi Chiu	P.S.Chiu
P.C.Chiu	Pei-Xi Chiu (= Pei-Shi Chiu)	P.S.Chiu
P.H.Chiu	Pei-Hsi Chiu	P.S.Chiu
H.Christ	Hermann Christ	Christ
Christ et Gies.	Hermann Christ et Karl Friedrich Georg Giesenhagen	Christ et Giesenh.
Clarke	Charles Baron Clarke	C.B.Clarke
Clarke et Bak.	Charles Baron Clarke et John Gilbert Baker	C.B.Clarke et Baker
Clausen	Robert Theodore Clausen	R.T.Clausen
Clayton	John Clayton	J.Clayton
Clem.	Simon de Rojas Clemente y Rubio	Clemente
Col.	William Colenso	Colenso
Colm.	Miguel Colmeiro	Colmeiro
Commerson	Philibert Commerson	Comm.
Cook	Orator Fuller Cook	O.F.Cook
Cop.	Edwin Bingham Copeland	Copel.
Copeland	Edwin Bingham Copeland	Copel.
Cord.	Eugène Jacob de Cordemoy	Cordem.
J.M.Coulter	John Merle Coulter	J.M.Coult.
Cuf.	Georg Cufodontis	Cufod.
Currey	Frederick Currey	Curr.
Czer.	Sergei Kirillovich Czerepanov	Czerep.
d'Alm.	Joseph Francis Raphael d'Almeida	d'Almeida
C.d'Orb.	Charles Victor Dessalines d'Orbigny	Orb.
Damaz.	Léonidas Botelho Damazio	Damazio
Dänik.	Albert Ulrich Däniker	Däniker
Dav.	George Edward Davenport	Davenp.
v.Deck.	Carl Claus von der Decken	Decken
I.Degen.	Irmgard Degener	I.Deg.

I.Degener	Irmgard Degener	**I.Deg.**
Degen.	Otto Degener	**O.Deg.**
Degener	Otto Degener	**O.Deg.**
de Licht.	Juana de Schafer Lichtenstein	**J.S.Licht.**
Demar.	Ferdinand Mathieu Hubert Demaret	**Demaret**
Don	David Don	**D.Don**
Donn.Smith	John Donnell Smith	**Donn.Sm.**
Dry.	Jonas Carlsson Dryander	**Dryand.**
Dup.	Louis Isidore Duperrey	**Duperrey**
Durand	Théophile Alexis Durand	**T.Durand**
Th.Durand	Théophile Alexis Durand	**T.Durand**
A.A.Eat.	Alvah Augustus Eaton	**A.A.Eaton**
Eat.	Daniel Cady Eaton	**D.C.Eaton**
Eat. et Dav.	Daniel Cady Eaton et George Edward Davenport	**D.C.Eaton et Davenp.**
D.C.Eat.	Daniel Cady Eaton	**D.C.Eaton**
Emys	[*sphalmate*] J.D. Enys	**Enys**
Espin.	Marcial Ramón Espinosa Bustos	**Espinosa**
Ett.	Constantin von Ettingshausen	**Ettingsh.**
O.Evans	Obed David Evans	**O.D.Evans**
Farwell	Oliver Atkins Farwell	**Farw.**
Fedch.	Boris Alexjewitsch Fedtschenko (Fedchenko)	**B.Fedtsch.**
O.Fedch.	Olga Alexandrowna Fedtschenko (Fedchenko)	**O.Fedtsch.**
Fedtsch.	Boris Alexjewitsch Fedtschenko	**B.Fedtsch.**
Fern.	Merritt Lyndon Fernald	**Fernald**
Ros.Fernandes	Rosette Mercedes Saraiva Batarda Fernandes	**R.Fern.**
Field. et Gardn.	Henry Barron Fielding et George Gardner	**Fielding et Gardner**
Filarszky	Nándor Filarszky	**Fil.**
Fisch. et Mey.	Friedrich Ernst Ludwig von Fischer et Carl Anton von Meyer	**Fisch. et C.A.Mey.**
Forbes	Henry Ogg Forbes	**H.O.Forbes**
Forsk.	Pehr (Peter) Forsskål	**Forssk.**
Forst.	Johann Georg Adam Forster	**G.Forst.**
Fosb.	Francis Raymond Fosberg	**Fosberg**
Fourn.	Eugène Pierre Nicolas Fournier	**E.Fourn.**
Fraser-Jenkins	Christopher Roy Fraser-Jenkins	**Fraser-Jenk.**
R.E.Fries	Robert Elias Fries	**R.E.Fr.**
Friesn.	Ray Clarence Friesner	**Friesner**
Fu	Shu-Hsia Fu	**S.H.Fu**
Fuchs	Leonhart Fuchs (Fuchsius)	**L.Fuchs**
Gal.	Henri Guillaume Galeotti	**Galeotti**
Garc.	Donato García	**D.García**

Gardn.	George Gardner	**Gardner**
Gaspar.	Guglielmo Gasparrini	**Gasp.**
Gaud.	Charles Gaudichaud-Beaupré	**Gaudich.**
Gepp	Anthony Gepp	**A.Gepp**
Gerr.	William Tyrer Gerrard	**Gerrard**
Ghatak	Jagadananda Ghatak	**J.Ghatak**
Ghosh	Santi Ranjan Ghosh	**S.R.Ghosh**
Gies.	Karl Friedrich Georg Giesenhagen	**Giesenh.**
Gilb.	Benjamin Davis Gilbert	**Gilbert**
E.F.Gilb.	Elizabeth Florence Gilbert	**E.F.Gilbert**
Glassm.	Sidney Frederick Glassman	**Glassman**
Gleas.	Henry Allan Gleason	**Gleason**
Gmel.	Johann Friedrich Gmelin	**J.F.Gmel.**
Gmel.	Samuel Gottlieb Gmelin	**S.G.Gmel.**
J.F.Gmelin	Johann Friedrich Gmelin	**J.F.Gmel.**
S.G.Gmelin	Samuel Gottlieb Gmelin	**S.G.Gmel.**
Goebel	Karl Immanuel Eberhard von Goebel	**K.I.Goebel**
Goepp.	Johann Heinrich Robert Göppert (Goeppert)	**Göpp.**
Golic.	Sergey Vladimirovich Golitsin (Golicin)	**Golitsin**
L.Gómez	Luis Diego Gómez	**L.D.Gómez**
Grant	Verne Edwin Grant	**V.E.Grant**
S.F.Gray	Samuel Frederick Gray	**Gray**
Greth.	David Frank Grether	**Grether**
Grev. et Hk.	Robert Kaye Greville et William Jackson Hooker	**Grev. et Hook.**
Gris.	August Heinrich Rudolf Grisebach	**Griseb.**
Grub.	Valery Ivanovich Grubov	**Grubov**
Guillaum.	André Guillaumin	**Guillaumin**
Guillemin	Jean Baptiste Antoine Guillemin	**Guill.**
Gupta	Kedar Mal Gupta	**K.M.Gupta**
v.Hall	Herman Christiaan van Hall	**H.C.Hall**
Ham.	Francis Hamilton (né Buchanan F.)	**Buch.-Ham.**
F.Ham.	Francis Hamilton (né Buchanan F.)	**Buch.-Ham.**
J.Hanst.	Johannes Ludwig Emil Robert von Hanstein	**Hanst.**
Hara	Hiroshi Hara	**H.Hara**
Harley	Winifred Jewell Harley	**W.J.Harley**
J.G.Hawkes	John Gregory Hawkes	**Hawkes**
Hay.	Bunzô Hayata	**Hayata**
H.B.K.	Friedrich Wilhelm Heinrich Alexander von Humboldt, Aimé Jacques Alexandre Bonpland et Karl Sigismund Kunth	**Humb., Bonpl. et Kunth**
H.B.W.	Friedrich Wilhelm Heinrich Alexander von Humboldt, Aimé Jacques Alexandre Bonpland ex Carl Ludwig von Willdenow	**Humb., Bonpl. ex Willd.**
Heller	Amos Arthur Heller	**A.Heller**

Henn.

Henn.	Elbert Hennipman	**Hennipman**
P.Herm.	Paul Hermann (Hermannus)	**Herm.**
Herz.	Theodor Carl Julius Herzog	**Herzog**
Hew.	Robert Heward	**Heward**
Heyw.	Vernon Hilton Heywood	**Heywood**
Hier.	Georg Hans Emmo Wolfang Hieronymus	**Hieron.**
Hiit.	Henrik Ilmari Augustus Hiitonen	**Hiitonen**
Hill.	Wilhelm Hillebrand	**Hillebr.**
J.Hill	John Hill	**Hill**
Hilleb.	Wilhelm Hillebrand	**Hillebr.**
Hitchc.	Edward Hitchcock	**E.Hitchc.**
Hk.	William Jackson Hooker	**Hook.**
Hk. et Arn.	William Jackson Hooker et George Arnott Walker Arnott	**Hook. et Arn.**
Hk. et Bak.	William Jackson Hooker et John Gilbert Baker	**Hook. et Baker**
Hk. et Grev.	William Jackson Hooker et Robert Kaye Greville	**Hook. et Grev.**
Hk.f.	Joseph Dalton Hooker	**Hook.f.**
Hollow.	John Ernest Holloway	**Holloway**
Hombr. et Jacq.	Jacques Bernard Hombron et Honoré Jacquinot	**Hombr. et Jacquinot**
J.D.Hook.	Joseph Dalton Hooker	**Hook.f.**
Hope	Charles William Webley Hope	**C.Hope**
Houlst.	John Houlston	**Houlston**
T.T.Hsieh	[*sphalmate*] Yin-Tang Hsieh	**Y.T.Hsieh**
Hsu	Ying-Pen Hsu	**Y.P.Hsu**
Hult.	Eric Oskar Gunnar Hultén	**Hultén**
Hunter	William Hunter	**W.Hunter**
Husnot	Pierre Tranquille Husnot	**Husn.**
Iwatsuki	Kunio Iwatsuki	**K.Iwats.**
Janchen	Erwin Emil Alfred Janchen	**Janch.**
J.Jansen	Johannes Theodorus Jansen	**J.T.Jansen**
P.Jansen	Pieter Jansen	**Jansen**
Jenm.	Georg Samuel Jenman	**Jenman**
Joaquin	Joaquin María de Barnola	**Barnola**
Johns.	James Yate Johnson	**J.Y.Johnson**
Joncheere	Gerardus Johannes Pieter de Joncheere	**de Jonch.**
Jones	Marcus Eugene Jones	**M.E.Jones**
Karst.	George Karsten	**G.Karst.**
Kato	Masahiro Kato	**M.Kato**
Keys.	Alexander Friedrich Michael Leberecht Arthur von Keyserling	**Keyserl.**

70

Kickx f.	Jean Kickx (filius)	**J.Kickx f.**
T.Kirk	Thomas Kirk	**Kirk**
Kitagawa	Masao Kitagawa	**Kitag.**
Kl.	Johann Friedrich Klotzsch	**Klotzsch**
Klf.	Georg Friedrich Kaulfuss	**Kaulf.**
Klob.-Alis.	Eugenija Nikolaevna Alissova-Klobukova	**Aliss.**
Klotz.	Johan Friedrich Klotzsch	**Klotzsch**
Knuth	Reinhard Gustav Paul Knuth	**R.Knuth**
Koch	Karl Heinrich Emil Koch	**K.Koch**
Koch	Wilhelm Daniel Joseph Koch	**W.D.J.Koch**
G.C.T.Koch	Christian Theodor Koch-Grünberg	**Koch-Grünb.**
W.Koch	Wilhelm Daniel Joseph Koch	**W.D.J.Koch**
Köhler	Erich Köhler	**Er.Köhler**
X.X.Kong	Hsian-Shiu Kung (= Xian-Xu Kong)	**H.S.Kung**
König	Johann Gerhard König	**J.König**
Kolhat.	G.G. Kolhatkar	**Kolh.**
Komar.	Vladimir Leontjevich Komarov	**Kom.**
C.Koss.	Constantin Constantinovich Kossinsky	**Kossinsky**
Kramer	Karl Ulrich Kramer	**K.U.Kramer**
Krecz.	Vitaly Ivanovich Kreczetowicz	**V.I.Krecz.**
Krieger	Walther Krieger	**W.Krieg.**
O.Ktze.	Carl Ernst Otto Kuntze	**Kuntze**
Kümm.	Jenö Belá Kümmerle	**Kümmerle**
Kunkel	Günther Kunkel	**G.Kunkel**
Kuo	Chen-Meng Kuo	**C.M.Kuo**
Kurata	Satoru Kurata	**Sa.Kurata**
Kze.	Gustav Kunze	**Kunze**
L'Hérit.	Charles-Louis L'Héritier de Brutelle	**L'Hér.**
Lab.	Jacques Julien Houtton de Labillardière	**Labill.**
Lac.	Charles Charmichael Lacaita	**Lacaita**
Lachm.	Jean Paul Lachman	**P.Lachm.**
Lag., García et Clem.	Mariano de Lagasca y Segura, Donato García et Simon de Rojas Clemente y Rubio	**Lag., D.García et Clemente**
Y.C.Lan	Young-Zhen (Young-Chen) Lan	**Y.Z.Lan**
A.Láng	Adolph Franz Láng	**Láng**
Larr.	Dámaso Antonio Larrañaga	**Larrañaga**
Laut.	Carl Adolf Georg Lauterbach	**Lauterb.**
Lawall.	André Gilles Célestin Lawalrée	**Lawalrée**
Lawson	George Lawson	**G.Lawson**
Le Grand	Antoine Legrand	**Legrand**
Lév.	Augustin Abel Hector Léveillé	**H.Lév.**
Léveillé	Augustin Abel Hector Léveillé	**H.Lév.**
Liljebl.	Samuel Liljeblad	**Lilj.**
A.Lima	Arturo Dárdano de Andrade Lima	**Andrade Lima**
Y.X. Ling	You-Xing Lin	**Y.X.Lin**

71

Linn.	Carolus Linnaeus (pater)	**L.**
W.J.Linton	William James Linton	**Linton**
Liou	Tchen-Ngo Liou	**T.N.Liou**
Lois.	Jean Louis Auguste Loiseleur-Deslongchamps	**Loisel.**
Losch	Ilse Losch	**I.Losch**
Loud.	John Claudius Loudon	**Loudon**
Löve	Åskell Löve	**Å.Löve**
R.Lowe	Richard Thomas Lowe	**Lowe**
Ludwig	Christian Gottlieb Ludwig	**Ludw.**
Macbr.	James Francis Macbride	**J.F.Macbr.**
MacKen	Mark John McKen (MacKen)	**McKen**
Mackenz.	Kennet Kent Mackenzie	**Mack.**
Mak.	Tomitarô Makino	**Makino**
Mäkinen	Yrjö Lauri Antero Mäkinen	**Y.Mäkinen**
X.Manetti	Saverio (Xaverius) Manetti	**Manetti**
Mann	Horace Mann	**H.Mann**
Marie-Vict.	Frère Marie Victorin	**Vict.**
Mart. et Gal.	Martin Martens et Henri Guillaume Galeotti	**M.Martens et Galeotti**
Martens	Martin Martens	**M.Martens**
Matthew	Charles Geekie Matthew	**C.G.Matthew**
Mayebara	Kanjiro Mayebara	**Mayeb.**
McCoy	Thomas Nevil McCoy	**T.N.McCoy**
Melzer	Helmut Melzer	**H.Melzer**
Melz.	Helmut Melzer	**H.Melzer**
Menz.	Archibald Menzies	**Menzies**
Merrill	Elmer Drew Merrill	**Merr.**
D.E. Meyer	Dieter Erich Meyer	**D.E.Mey.**
M.Mich.	Marc Micheli	**Micheli**
P.Mich.	Pier'Antonio Micheli (Michelius)	**P.Micheli**
Mikheladze	Irakly Alexandrovich Mikheladze	**Mikhel.**
J.Milde	(Carl August) Julius Milde	**Milde**
H.Miller	Harvey Alfred Miller	**H.A.Mill.**
Min.	Nicolai Aleksandrovich Miniaev	**Miniaev**
Miou	[*sphalmate*] Ru-Huai Miau	**R.H.Miau**
D.Mitch.	David Searle Mitchell	**D.S.Mitch.**
X.L.Mo	Sin-Li Mo	**S.L.Mo**
Moore	Thomas Moore	**T.Moore**
D.Moore	David Moore	**Moore**
Moore et Houlst.	Thomas Moore et John Houlston	**T.Moore et Houlston**
Mori	Tamezô Mori	**T.Mori**
Moritz.	Alexander Moritzi	**Moritzi**
Morton	Conrad Vernon Morton	**C.V.Morton**
C.Morton	Conrad Vernon Morton	**C.V.Morton**

Moxl.	George Loucks Moxley	Moxley
F.v.Muell.	Ferdinand Jacob Heinrich von Mueller	F.Muell.
K.Muell.berol.	Karl Müller (Berolinensis)	Müll.Berol.
Müll.	Otto Friedrich Müller	O.F.Müll.
K.Müll.	Johann Karl August Müller [Halensis]	Müll.Hal.
K.Müll.hal.	Johann Karl August Müller [Halensis]	Müll.Hal.
Müller Halle	Johann Karl August Müller [Halensis]	Müll.Hal.
J.A.Murray	Johan Andreas Murray	Murray
Nair	Narayana Chandrashekharan Nair	N.C.Nair
Nakam.	Takehisa Nakamura	T.Nakam.
Namegata	Tomitaro Namegata	Nameg.
Nardi	Enio Nardi	E.Nardi
B.Nayar	Bala Krishnan Nayar	B.K.Nayar
E.E.Nels.	Elias Emanuel Nelson	E.E.Nelson
Nelson	Aven Nelson	A.Nelson
Nichols.	George Nicholson	G.Nicholson
G.Nichols.	George Nicholson	G.Nicholson
Nish.	Makoto Nishida	M.Nishida
Nishida	Makoto Nishida	M.Nishida
Noroña	Francisco Noroña (Noronha)	Noronha
Nyl.	Fredrik Nylander	F.Nyl.
Nyland.	Fredrik Nylander	F.Nyl.
Nym.	Carl Fredrik Nyman	Nyman
Oed.	George Christian Oeder	Oeder
M.Ogata	Masasuke Ogata	Ogata
Oliver	Daniel Oliver	Oliv.
Ort.	Casimiro Gómez (de) Ortega	Ortega
C.Page	Christopher Nigel Page	C.N.Page
Palibin	Ivan Vladimirovich Palibin	Palib.
Palmer	Thomas Chalkley Palmer	T.C.Palmer
Panigr.	Gopinath Panigrahi	Panigrahi
Pant	Divya Darshan Pant	D.D.Pant
Pappe et Raws.	Karl Wilhelm Ludwig Pappe et Rawson William Rawson	Pappe et Rawson
Parker	Charles Sandbach Parker	C. Parker
Parm.	Paul Evariste Parmantier	P.Parm.
A.Parm.	Antoine Auguste Parmentier	Parm.
Pasch.	Adolf Pascher	Pascher
Paterson	Robert H. Paterson	R.H.Paterson
N.Pavl.	Nikolai Vasilievich Pavlov	Pavlov
Pavl.	Nikolai Vasilievich Pavlov	Pavlov
PB.	Ambroise Marie François Joseph Palisot de Beauvois	P.Beauv.

Pérez	Enrique Pérez Arbeláez	**Pérez Arbel.**
Pérez Arb.	Enrique Pérez Arbeláez	**Pérez Arbel.**
Perkins	Janet Russel Perkins	**J.R.Perkins**
Perrottet	George Samuel Perrottet	**Perr.**
A.Peter	Albert Peter	**Peter**
Pfitz.	Ernst Hugo Heinrich Pfitzer	**Pfitzer**
Pichi	Rodolfo Emilio Giuseppe Pichi Sermolli	**Pic.Serm.**
Pic.Ser.	Rodolfo Emilio Giuseppe Pichi Sermolli	**Pic.Serm.**
Pichi-Serm.	Rodolfo Emilio Giuseppe Pichi Sermolli	**Pic.Serm.**
M.Pinto	Manuel Cabral de Resende-Pinto	**Res.-Pinto**
Pitt.	Henri François Pittier	**Pittier**
Poirault	Marie Henri Georges Poirault	**G.Poirault**
Porsild	Alf Erling Porsild	**A.E.Porsild**
M.Porsild	Morton Pedersen Porsild	**Porsild**
Port.	Franz von Portenschlag-Ledermayer	**Port.-Led.**
Portenschl.	Franz von Portenschlag-Ledermayer	**Port.-Led.**
Pospichal	Eduard Pospichal	**Posp.**
Pr.	Carl Bořivoj Presl	**C.Presl**
Presl	Carl Bořivoj Presl	**C.Presl**
J.S.Presl	Jan Svatopluk Presl	**J.Presl**
K.Presl	Carl (Karl) Bořivoj Presl	**C.Presl**
M.Price	Michael Greene Price	**M.G.Price**
Pritch.	Stephen F. Pritchard	**S.Pritch.**
Pylaie	Auguste Jean Marie Bachelot de la Pylaie	**Bach.Pyl.**
Raimann	Rudolf Raimann	**Raim.**
Raws.	Rawson William Rawson	**Rawson**
Reching.	Karl Heinz Rechinger	**Rech.f.**
Reed	Clyde Franklin Reed	**C.F.Reed**
C.Reed	Clyde Franklin Reed	**C.F.Reed**
Rgl.	Eduard von Regel	**Regel**
Reich.	Johann Jakob Reichard	**Reichard**
Reich.	Heinrich Wilhelm Reichardt	**Reichardt**
Reich.fil.	Heinrich Gustav Reichenbach	**Rchb.f.**
Reichb.f.	Heinrich Gustav Reichenbach	**Rchb.f.**
Rémy	Ezechiel Jules Rémy	**J.Rémy**
Rey-Pailhade	Constantin de Rey-Pailhade	**Rey-Pailh.**
Richards.	John Richardson	**Richardson**
Al.Richt.	Vincenz Aladár Richter	**V.A.Richt.**
A.Richter	Vincenz Aladár Richter	**V.A.Richt.**
J.Riley	John Riley	**Riley**
Robins. et Fernald	Benjamin Lincoln Robinson et Merritt Lyndon Fernald	**B.L.Rob. et Fernald**
B.L.Robins.	Benjamin Lincoln Robinson	**B.L.Rob.**
B.L.Robinson	Benjamin Lincoln Robinson	**B.L.Rob.**
J.Robinson	John Robinson	**J.Rob.**

W.J.Robinson	Winifred Josephine Robinson	**W.J.Rob.**
Roivainen	Heikki Roivainen	**Roiv.**
Rojas	Nicolás Rojas Acosta	**Rojas Acosta**
Ros.	Eduard Rosenstock	**Rosenst.**
J.Rose	Joseph Nelson Rose	**Rose**
Rosend.	Henrik Viktor Rosendahl	**H.Rosend.**
Rosenstock	Eduard Rosenstock	**Rosenst.**
Ross	Hermann Ross	**H.Ross**
Rovir.	José N. Rovirosa	**Rovirosa**
Roxas	Simon de Rojas (Roxas) Clemente y Rubio	**Clemente**
R.R.S.	Ralph Randles Stewart	**R.R.Stewart**
R. & S.	Johan Jacob Roemer et Joseph August Schultes	**Roem. et Schult.**
Rudm.Brown	Robert Neal Rudmose Brown	**R.N.R.Br.**
Sadebeck	Richard Emil Benjamin Sadebeck	**Sadeb.**
Sadl.	Joseph Sadler	**Sadler**
Jos.Sadler	Joseph Sadler	**Sadler**
Sal.	Carl (E.) Salomon	**Salomon**
A.J.Sampaio	Alberto José de Sampaio	**A.Samp.**
G.Sampaio	Gonçalo António da Silva Ferreira Sampaio	**Samp.**
Sav.	Jule César Lelorgne de Savigny	**Savigny**
Savat.	Paul Amadée Ludovic Savatier	**Sav.**
Schaffn.	Wilhelm Schaffner	**W.Schaffn.**
Schauer	Sebastian Schauer	**S.Schauer**
Schlecht.	Diederich Franz Leonhard von Schlechtendal	**Schltdl.**
Schlechtend.	Diederich Franz Leonhard von Schlechtendal	**Schltdl.**
Schmid	Rupertus Schmid	**Rup. Schmid**
Schmid.	Casimir Christoph Schmidel	**Schmidel**
Geo.Schneid.	George Schneider	**G.Schneid.**
U.Schneider	Ulrike Schneider	**U.Schneid.**
Scholtz	Johann Eduard Heinrich Scholtz	**H.Scholtz**
Schum.	Christian Friedrich Schumacher	**Schumach.**
Schum. et Laut.	Karl Moritz Schumann et Carl Adolf Georg Lauterbach	**K.Schum. et Lauterb.**
Schwarz	Otto Karl Anton Schwarz	**O.Schwarz**
Scott	John Scott	**J.Scott**
Scott	Robert Robinson Scott	**R.R.Scott**
Jo.Scott	John Scott	**J.Scott**
Seat.	Henry Eliason Seaton	**Seaton**
Séguier	Jean François Séguier	**Ség.**
Sen	Udayananda Sen	**U.Sen**
F.Seymour	Frank Conkling Seymour	**F.Seym.**
Shieh	Wang-Chueng Shieh	**W.C.Shieh**
Shing	Kung-Hsia Shing	**K.H.Shing**
Shuttleworth	Robert James Shuttleworth	**Shuttlew.**

Sipl.	Vladimir N. Siplivinsky	**Sipliv.**
Slosson	Margaret Slosson	**Sloss.**
John Sm.	John Smith	**J.Sm.**
J.D.Sm.	John Donnell Smith	**Donn.Sm.**
Smith	James Edward Smith	**Sm.**
A.C.Smith	Albert Charles Smith	**A.C.Sm.**
A.Reid Smith	Alan Reid Smith	**A.R.Sm.**
D.M.Smith	Dale Metz Smith	**D.M.Sm.**
J.Smith	John Smith	**J.Sm.**
Sod.	Luis Sodiro	**Sodiro**
Sota	Elias Ramon de la Sota	**de la Sota**
Spr.	Kurt Sprengel	**Spreng.**
Srivastava	Gopal Krishna Srivastava	**G.K.Srivast.**
St.-Hil.	Auguste François César Prouvençal de Saint-Hilaire	**A.St.-Hil.**
St.John	Harold St.John	**H.St.John**
E.St.John	Edward Porter St.John	**E.P.St.John**
Stewart	Ralph Randles Stewart	**R.R.Stewart**
R.Stewart	Ralph Randles Stewart	**R.R.Stewart**
Stone	Benjamin Clemens Masterman Stone	**B.C.Stone**
Stremp.	Johannes Karl Friedrich Strempel	**Strempel**
Sturm	Johann Wilhelm Sturm	**J.W.Sturm**
Suesseng.	Karl Suessenguth	**Suess.**
Sumn.	Georgij Prokopievic Sumnevicz	**Sumnev.**
Suzuki	Hyoji Suzuki	**H.Suzuki**
Svens.	Henry Knute Svenson	**Svenson**
Swartz	Olof Swartz	**Sw.**
Szysz.	Ignaz von Szyszylowicz	**Szyszyl.**
Tag.	Motozoi Tagawa	**Tagawa**
Takeuki	Masayuki Takeuki	**M.Takeuki**
Tard.	Marie-Laure Tardieu-Blot	**Tardieu**
Tardieu-Blot	Marie-Laure Tardieu-Blot	**Tardieu**
D.Tate	Donald Eugene Tate	**D.E.Tate**
T.M.C.Tayl.	Thomas Mayne Cunninghame Taylor	**T.M.C.Taylor**
A.Terracc.	Achille Terracciano	**A.Terrac.**
Thbg.	Carl Peter Thunberg	**Thunb.**
R.Thomps.	Rufus Henney Thompson	**R.H.Thomps.**
Thomson	George Malcolm Thomson	**G.M.Thomson**
Thwait.	George Henry Kendrick Thwaites	**Thwaites**
Tind.	Mary Douglas Tindale	**Tindale**
Trev.	Vittore Trevisan de Saint-Léon	**Trevis.**
W.Troll	Wilhelm Troll	**Troll**
Tryon	Rolla Milton Tryon	**R.M.Tryon**
Tuckerm.	Edward Tuckerman	**Tuck.**
Tzvel.	Nikolai Nikolaievich Tzvelev	**Tzvelev**

Und.	Lucien Marcus Underwood	**Underw.**
Und. et Cook	Lucien Marcus Underwood et Orator Fuller Cook	**Underw. et O.F.Cook**
Und. et Maxon	Lucien Marcus Underwood et William Ralph Maxon	**Underw. et Maxon**
Urban	Ignatz Urban	**Urb.**
Usteri	Alfred Usteri	**A.Usteri**
v.A.v.R.	Cornelis Rugier Willem Karel van Alderwerelt van Rosenburgh	**Alderw.**
v.Ald.v.Ros.	Cornelis Rugier Willem Karel van Alderwerelt van Rosenburgh	**Alderw.**
v.d.B.	Roelof Benjamin van den Bosch	**Bosch**
v.d.Bosch	Roelof Benjamin van den Bosch	**Bosch**
Van Hoorebeke	Charles Joseph Van Hoorebeke	**Van Hooreb.**
Vauch.	Jean Pierre Étienne Vaucher	**Vaucher**
Venten.	Étienne Pierre Ventenat	**Vent.**
Versch.	Ambroise Colette Alexandre Verschaffelt	**Verschaff.**
Victor.	Frère Marie Victorin	**Vict.**
Vidal	Louis Vidal	**L.Vidal**
Vidal	Sebastian Vidal y Soler	**S.Vidal**
Vitm.	Fulgenzio Vitman	**Vitman**
Vogler	Johann (Philipp) Andreas Vogler	**J.A.Vogler**
J.P.Vogler	Johann (Philipp) Andreas Vogler	**J.A.Vogler**
J.Voigt	Joachim Otto Voigt	**Voigt**
Vollmann	Franz Vollmann	**Vollm.**
A.Wade	Arthur Edwin Wade	**A.E.Wade**
Wagner	Warren Herbert Wagner Jr.	**W.H.Wagner**
W.Wagner	Warren Herbert Wagner Jr.	**W.H.Wagner**
Wakef.	Norman Arthur Wakefield	**N.A.Wakef.**
T.Walker	Trevor George Walker	**T.G.Walker**
Wallich	Nathaniel Wallich	**Wall.**
Walt.	Thomas Walter	**Walter**
Wang	Chu-Hao Wang	**Chu H.Wang**
C.H.Wang	Chu-Hao Wang	**Chu H.Wang**
C.R.Wang	Chung-Ren Wang = Zhong-Ren Wang	**Z.R.Wang**
Z.B.Wang	[sphalmate] Zhong-Ren Wang	**Z.R.Wang**
Wannt.	Hans-Erik Wanntorp	**Wanntorp**
E.Warb.	Edmund Frederic Warburg	**E.F.Warb.**
H.Wats.	Hewett Cottrell Watson	**H.C.Watson**
Watson	Hewett Cottrell Watson	**H.C.Watson**
D.Watt	David Allan Poe Watt	**Watt**
P.Webb	Philip Barker Webb	**Webb**
Webb et Berth.	Philip Barker Webb et Sabin Berthelot	**Webb et Berthel.**
G.H.Weber	Georg Heinrich Weber	**Weber**

F.W.Weiss	Friedrich Wilhelm Weiss	**Weiss**
Whitwell	William Whitwell	**Whitw.**
Wigg.	Fridrich Hindrich Wiggers	**F.H.Wigg.**
J.Wilce	Joan Hubbell Wilce	**J.H.Wilce**
Wilcz.	Ernst Wilczek	**Wilczek**
Willem.	Pierre Rémi François de Paule Willemet	**P.Willemet**
K.Wilson	Kenneth Allen Wilson	**K.A.Wilson**
Wind.	Paulo Günther Windisch	**P.G.Windisch**
E.Winslow	Evelyn James Winslow	**E.J.Winslow**
Wither.	William Withering	**With.**
Wollast.	Georg Buchanan Wollaston	**Woll.**
A.Wood	Alphonso (W.) Wood	**A.W.Wood**
Woynar	Heinrich Karl Woynar	**Woyn.**
Wright	Charles Wright	**C.Wright**
Wu	Yin-Chan Wu	**Y.C.Wu**
Wulf.	Franz Xavier von Wulfen	**Wulfen**
O.Wünsche	Friedrich Otto Wünsche	**Wünsche**
Y.T.Xie	Yin-Tang Hsieh (Xie)	**Y.T.Hsieh**
Yabe	Yoshitaka Yabe	**Y.Yabe**
Chen Y.Yang	Chun-Yu Yang	**C.Y.Yang**
Yao	Guan-Hu Yao	**G.H.Yao**
L.Yates	Lorenzo Gordin Yates	**Yates**
C.Y.Yaug	[*sphalmate*] Chan-You Yang	**Chang Y. Yang**
Yuncker	Truman George Yuncker	**Yunck.**
Zamora	Prescillano Martinez Zamora	**P.M.Zamora**
Zenk.	Jonathan Carl Zenker	**Zenker**
W.Zimmerm.	Walter Max Zimmermann	**W.Zimm.**
Zippel.	Alexander Zippelius	**Zipp.**
Y.C.Zong	[*sphalmate*] Ye-Cong Zhong	**Y.C.Zhong**

Printed in Italy by Litografia Europa, La Spezia

July 1996